November 1994.
Johannesburg.

Sandy & Butch,

Wishing you a visit to remember in our New South Africa!

All the best,

Douglas & Barbara..

NATIONAL
PARKS
of
SOUTH AFRICA

NATIONAL PARKS
of
SOUTH AFRICA

PHOTOGRAPHY
ANTHONY BANNISTER
TEXT
BRENDAN RYAN

STRUIK

DEDICATIONS

BRENDAN RYAN
To my wife, Ingrid

ANTHONY BANNISTER
To Barbara, David, Jemima, Patrick, Andrew and Susan

All photographs in this book supplied by
ABPL (Anthony Bannister Photo Library)
PO Box 11, Lanseria, 1748 South Africa
Tel: (011) 701-3000 Fax (011) 701-3003

CONSULTANTS

DR ANTHONY HALL-MARTIN, NATIONAL PARKS BOARD
DR GUS MILLS, NATIONAL PARKS BOARD
DR JOHN HANKS, SOUTHERN AFRICAN NATURE FOUNDATION
GARY MAY

Struik Publishers
(a member of The Struik Group (Pty) Ltd)
Cornelis Struik House, 80 McKenzie Street
Cape Town 8001

House editor: Lindsay Norman
Editor: John Comrie-Greig
Designer: Kevin Shenton

Reproduction by Unifoto (Pty) Ltd
Printed and bound by South China Printing Co., Hong Kong

ISBN 1 86825 337 6

HALF TITLE: *Yellow mongooses, Karoo National Park.*
TITLE PAGE: *Mountain View Lookout, Karoo National Park.*
RIGHT: *Olifants River, Kruger National Park.*

CONTENTS

ACKNOWLEDGEMENTS

ANTHONY BANNISTER

In the 10 years since I photographed the earlier version of *National Parks of South Africa* (Struik, 1983) the number of national parks has grown from 12 then to a total of 17 today, with several more waiting in the wings. In the interim the National Parks Board continues to evolve into a far more streamlined organization, outward looking, largely self-financing, yet providing affordable and exciting experience for all South Africans. There can be no doubt that South Africa has the most diverse and best managed system of national parks in all Africa.

Producing a collection of photographs that accurately reflect the unique spirit of 17 national parks spread across the greatness of South Africa, required no small expenditure of time and money. Fortunately, a number of South African sponsors contributed generously to offset many of the costs incurred in providing the photography for this book and I would particularly like to thank the following:

Richards Bay Minerals and Palabora Mining Company for their valuable contribution towards costs; Frank & Hirsch Ltd for professional servicing of my extensive set of Nikon photographic equipment; Beith Photographic Laboratory of Johannesburg for E6 processing to a standard equal to the best in the world; and finally, Speedair of Lanseria for professional air charter that made possible much of the aerial photography.

To the National Parks Board and its officials I am especially indebted for their invaluable contribution towards this book. The Board assisted me without hesitation in making its facilities available. Many park wardens, rangers and researchers readily gave up their personal time (often entire weekends) to assist. Thank you all. To Dr 'Robbie' Robinson, Chief Executive Director of the National Parks Board, I give my thanks for his personal encouragement of this project.

And finally, thanks to Barbara whose caring for our home and family during my all too frequent absences always makes returning a delight. I thank my older children Andrew and Susan for their cheerful assistance and great company on several visits to the national parks; and last but not least my secretary Sarah Peacock for her hard work and considerable patience.

BRENDAN RYAN

The author would like to acknowledge the help and guidance given to him by the wardens, rangers, researchers and information officers in all the national parks who provided essential background information, answered his innumerable questions and who were prepared, when asked to do so, to give their personal opinion on many important issues.

LEFT: *A herd of gemsbok gallop across a pan in the Kalahari Gemsbok National Park.*
OVERLEAF: *This view of the Orange River Gorge can be seen from the Oranjekom Lookout in the Augrabies Falls National Park.*

FOREWORD

It is indeed a great honour and a pleasure to be asked to contribute the foreword to this splendid book. National Parks in general, and in South Africa in particular, are being scrutinized and may in fact be threatened if we do not address the situation in a responsible manner. The publication of this book therefore comes at an opportune time.

On 2 February 1990, the State President of the Republic of South Africa, Mr F W de Klerk made his historic speech unbanning certain organizations, promising the release of political prisoners and launching the Republic of South Africa on the path of a negotiated transition to a new political system or a 'New South Africa' as he termed it. Since then, there has been much rhetoric and discussion on what form this new system should take. The constitutional debate is focused increasingly on the degree of devolution of government functions. The proposals in the media range from strong centralized government to a loose federation made up of several regions, with emphasis on local government.

For our national parks to be meaningful in the 'New South Africa' the Board will have to reposition itself in such a way as to be acceptable to the new political leaders and become less reliant on government funding. The organization will have to become more business-orientated while never losing sight of the original stewardship of preserving national parks for this and future generations in as natural a state as possible.

South Africa is blessed with one of the earth's richest and unique wildlife heritages. The diversity of climate, topography, ecosystems, animals and plant life, has made South Africa the focal point for travellers from all over the world. The variety and scenic beauty of South Africa's astonishing natural heritage has in part been protected in a system of national parks. This system is by no means complete but it does protect a wide spectrum of our natural resources, some of which cannot be found anywhere else on earth. To manage these national natural assets is indeed a great privilege but it also has an awesome responsibility attached to it. National parks must be managed in such a way that their integrity is not compromised for future generations and it is therefore important that their value must be clearly eluci-

Dr G. A. Robinson

dated, both economically and spiritually, to the new electorate of South Africa.

The mission of the National Parks Board is to establish a system of national parks representative of South Africa's important ecosystems and unique natural features and to conserve and manage them in such a way that they will be preserved for all time in their natural state for the benefit and inspiration of present and future generations of all South Africans.

My own vision is for our national parks to become the pride and joy of every citizen of South Africa and it is with these objectives in mind that I believe that this book will make a very significant impact.

The author, Brendan Ryan, has travelled many thousands of kilometres to all the areas included in this book. The care and attention to detail that has gone into the writing of it is indicative of his deep love and respect for our national parks. His professional journalistic talent has brought the issues that management is facing to the reader's attention whilst emphasizing the beauty and charm of our priceless wildlife heritage.

The photographer, Anthony Bannister, whom I have known for almost twenty-five years, is without doubt one of South Africa's leading wildlife photographers and his superb photographs have stimulated interest in our national parks all over the world. We have debated the necessity of publications of this sort and it is indeed gratifying to see Anthony's ideas coming to fruition. It is, therefore, my pleasure to thank these two individuals and to congratulate them on this magnificent and valuable publication.

I would also like to express my heartfelt appreciation to Struik Publishers and all their consultants and members of their staff for producing a book of this calibre. Not only will it contribute to the enjoyment of our visitors in stimulating interest, but it will also serve as a precious memento of a visit to these sacrosanct places.

Dr G. A. ROBINSON
Chief Executive Director
NATIONAL PARKS BOARD, PRETORIA

INTRODUCTION

An African fish eagle and its prey of barbel. It is capable of catching fish that weigh up to two kilograms.

The world has been turned upside-down in recent years with a number of previously unthinkable political events taking place as, for example, the unification of East and West Germany, the disintegration of the Soviet Union and the spurning of communism by many countries which formerly espoused that ideology.

Africa in general and South Africa in particular have not escaped the effects of these renewed 'winds of change' which have reshaped both politics and nature conservation. The end of apartheid and the radical re-assessment of generally accepted conservation policies have combined to alter for ever the world of the National Parks Board of South Africa – the statutory body responsible for running 16 national parks and one national lake area in broad accordance with the guidelines and the philosophy of the World Conservation Union (I.U.C.N.) which defines a national park in the following terms: 'A National Park is a relatively large, outstanding natural area managed by a nationally-recognized authority to protect the ecological integrity of one or more ecosystems for this and future generations and to eliminate any exploitation or intensive occupation of the area and to provide a foundation for spiritual, scientific, educational and tourism opportunities.'

The history of the National Parks Board shows that forceful and far-sighted men have made decisive impacts at critical periods during the course of its development. One of these men was Lieutenant-Colonel James Stevenson-Hamilton who was the first warden of the Kruger National Park. Another such man, Dr Gilbert Adrian 'Robbie' Robinson, is now at the helm of the organization and the changes he has already made are radical enough to warrant the same nickname which was given to Stevenson-Hamilton – 'Skukuza', which means 'he who turns everything upside-down'.

LEFT: *Waterbuck will readily take to water as a sanctuary from predators. They may even submerge their bodies, leaving only their nostrils out of the water.*

Since he took over as Chief Executive Director in 1991 he has set about changing both the shape of the Parks Board and certain outmoded aspects of its staff's 'corporate culture'. His driving motivation is the realization that certain key strategies have to be adopted if the Parks Board is to be able to continue with its mission in the new political system.

Policies which were previously unthinkable are now being introduced into Parks Board operations: these include affirmative-action programmes, equal employment opportunities and a non-discriminatory wage and salary structure. The organization itself has been streamlined through retrenchments and a modified management system that is geared for greater efficiency.

The frequently negative attitude of some of the Parks Board staff to outsiders has been challenged, to change it into one of greater openness towards the visitors and tourists on whom the future success and viability of South Africa's parks will depend.

Robinson says he, personally, is not the driving force behind this transformation: it results rather from the general and sweeping changes made in South Africa with the removal of apartheid and the move towards establishment of the 'New South Africa'. 'The State President, F.W. de Klerk, changed the *status quo* in February 1990 when he announced the various measures aimed at bringing about a new, democratic, South Africa. All I am doing is positioning my organization correctly in terms of this new situation', comments Robinson.

Changing an organization's structure and 'culture' is hard enough but Robinson is also confronting other critical issues simultaneously. The first is how the National Parks Board will come to terms with and implement the relatively new conservation credo of sustainable utilization, which can be translated into layman's terms as, 'if conservation pays, it stays'.

Related to this is the question of the long-term financial survival of the National Parks Board in a future where government funding for conservation must inevitably drop because of the other crushing priorities such as housing and education.

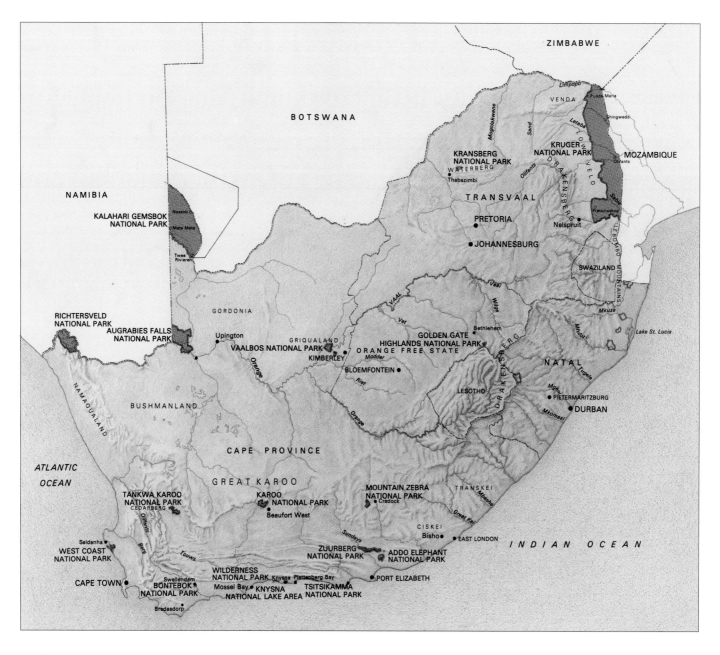

The National Parks Board has to become self-sufficient financially while at the same time it must keep accommodation rates affordable so that its parks do not become the preserve only of affluent international tourists and the wealthier segment of South African society. The Board must also have the financial strength to continue to expand the national park system to ensure that representative examples of all the ecosystems found in South Africa are adequately conserved.

One of the solutions may lie in the 'contractual' national park concept, where ownership of land remains in private hands but is managed by the Parks Board as a national park. A pioneering development of this type was the agreement reached in the Richtersveld which allowed pastoralists to continue to graze their livestock inside the boundaries of the Richtersveld National Park (see page 35).

When the Richtersveld agreement was signed in July 1991 it indicated to South Africa that the National Parks Board was serious about the way it intended doing business in the future and it changed the views of many of its most severe critics, which were that it was a racist blinkered bureaucracy doing things strictly by the rules even if the rules were clearly wrong. The proclamation of the Groenkloof National Park in Pretoria where the Parks Board's headquarters are situated is a case in point. The National Parks Act prevents the Board from spending money on construction outside a proclaimed national park so, in order to build its offices in Pretoria, the 6,76-hectare site was

Mist shrouds a bontebok in the Bontebok National Park which was founded specifically to protect this species.

declared a national park. 'That was ridiculous. Rather than maintain the pretence that Groenkloof is a national park, the rules should have been changed', Robinson points out.

The changes he has made are sweeping and result not only from the new political and social system now emerging in South Africa, but from his own previously experienced frustrations within the organization to which he has devoted his entire working career since joining it as the first warden of the Tsitsikamma Coastal National Park in 1966.

He cut the numbers of senior managerial staff at Parks Board headquarters from a top-heavy ten to five, keeping only one of the previous executives – Dr Salomon Joubert who is now Executive Director: Kruger National Park. He followed this with a programme of decentralization, transferring power from Parks Board headquarters to the various park wardens. Previously, several functions in a national park such as management of rest-camps, were reported on directly to the Pretoria

headquarters of the Parks Board without reference to the park warden. This system created conflict and confusion which has been removed now that the wardens are responsible for everything that happens in their parks.

That, perhaps, was the easiest part of his self-imposed task, because simultaneously Robinson set about changing the 'corporate culture' of the organization which had developed over the 65-odd years following the introduction of the National Parks Act in 1926. Amongst the N.G.O.s (non-governmental conservation organizations) it was known as the 'gardeners of Eden' syndrome, in which some Parks Board staff appeared to feel that their own interests came first and acted in a high-handed manner. Stories of rudeness to visitors in certain parks abounded, with one warden of another park experiencing it personally on a visit to the Kruger Park when he travelled incognito. Attitudes and service changed dramatically once he revealed who he was.

Watched by her almost fully grown cubs, a cheetah drags her kill into cover. Cheetahs are frequently forced off their kills by scavenging lion and hyena.

Comments Robinson: 'Visitors were seen as a bit of burden in terms of their presence in the parks and some staff didn't welcome them, they tolerated them. What I'm trying to get them to realize is that we are running a business and the visitors are our clients upon whom our future depends.'

The National Parks Board acquired 'international status' for all its rest-camps in the early 1980s when a loophole in apartheid legislation was introduced to allow hotels to become multi-racial. The Board, however, never actively advertised the fact. This attitude has now changed and the Parks Board now makes it clear that people of all races are welcome in all parks.

The previous negative attitude to visitors also manifested itself in office, shop and entrance-gate hours which were de-signed to suit the staff, not the visitors. Gate hours were a cause of particular frustration with the entrance gates to the Kruger Park, for example, closing over lunchtime and leaving cars full

of visitors to wait in the Lowveld heat for the gatekeepers to return. Robinson has extensively revised the hours of business to remedy this situation, and offices and entrance gates are now open throughout the day.

He has also introduced a non-discriminatory wage and salary structure and has gone further than any other South African official nature conservation body in the appointment of blacks to top positions. Ben Mokoatle was appointed Director of Human Resources, making him one of the five executive direc-tors who run the organization. The first black warden took up his post early in 1992 at the Zuurberg National Park and other suitably qualified members of race groups other than white are being recruited to the Board's communications department.

Robinson makes no bones about the fact that placing properly qualified blacks in senior positions in the organization is one of his objectives. 'If I have to choose between equally qualified black and white candidates for a post, I'll choose the black.' Such actions were unthinkable, and in some cases illegal, in terms of prevailing attitudes and laws only a few years ago and changes of this nature must have caused resentment in the organization.

Robinson concedes this but points out that the tremendous social and political changes sweeping the country have altered people's attitudes: this has helped him greatly in implementing his reforms.

Another two of Robinson's aims are to ensure the future financial self-sufficiency of the National Parks Board and to ensure its acceptance in the 'New South Africa'. This essentially means including the black majority in the conservation cause which in the past has essentially been the domain of the white minority. Both of these issues are inextricably intertwined.

For the year ending in March 1991 the Parks Board had a total income of R109-million of which R46,4-million came from accommodation and R33,1-million came through Government grants. Speculation is rife in environmental circles that the Government has told the National Parks Board the grants are going to be withdrawn. Robinson denies this but comments, 'Things are clearly going to get more stringent. The government of the "New South Africa" is not going to look too sympathetically on the financial requirements of my organization when it has such pressing concerns as housing, education and health to attend to. This country is in dire straits.'

Robinson plans to have the government subsidy phased out voluntarily within three years. The major contributor to earnings would then be accommodation and visitor services such as trails, curio shop sales and the like. Robinson says he needs to generate a real annual profit margin of about five per cent. That means a 20 per cent return on investment if South Africa's inflation rate keeps running at around 15 per cent per annum.

Depending on future trends in the inflation rate – and not even the optimists envisage that it will stay below 10 per cent – there will be a steady upward pressure on tariffs. The Parks Board, however, is caught in the cleft stick of trying to make its facilities affordable and accessible to all. Robinson has two answers to this. The first is that the parks offer a range of accommodation facilities, from ordinary camp-sites through to luxury private rest-camps; this presents a spread of prices that virtually anyone can afford. The second is Robinson's definition of his client base: the average South African who wants to visit the national parks. He points out that many South Africans are not interested in national parks or do not have sufficient disposable income to go on any kind of holiday. He has to keep the parks affordable for those who are interested, and he has to expand their numbers.

'We want to attract the people who are genuinely interested in the national parks and we plan to increase their numbers

through education. I do not see the national parks as recreational areas for people who are denied facilities such as playing-fields in the cities.'

The millions of impoverished neighbours living around some of the national parks are a different matter entirely and Robinson accepts that the Parks Board has to go out of its way to provide benefits for these people. At the same time, however, he must convince them of the need for the continued existence of the national parks system. Nevertheless, he again sees education as the most important means of achieving this and to this end Kruger Park staff have started a series of courses and programmes to spread the conservation message to opinion-makers and schoolchildren from the surrounding communities.

There are a number of strategic benefits that flow from the National Parks Board being financially independent. If it is not a drain on resources this could make a big difference in the attitude of a new majority government to the National Parks Board. It may also mean that the Board could plan its activities with confidence, unlike some other sister conservation organizations such as Zimbabwe's Department of National Parks and

A lion is perfectly camouflaged by the dry, yellow winter grass of the Kalahari Gemsbok National Park.

Sunset in the Kalahari Gemsbok National Park is the signal for many of its nocturnal animals to come out of the burrows and holes where they escape the burning heat of the day.

Wild Life Management which is a government department. All of the funds it raises flow to Zimbabwe's Central Treasury. The Department is allocated an annual budget by the Government but in recent years this has been cut annually as Zimbabwe's economy has deteriorated and demands from all sectors for Government funding have risen.

The attitude towards conservation of a future government dominated by what is, at the moment, the main black political party – the African National Congress (A.N.C.) – is largely unknown at this stage. However, the A.N.C. and other black political organizations such as the Pan-Africanist Congress (P.A.C.) and the Inkatha Freedom Party (I.F.P.) have published position papers on the issue of conservation which have proved to be generally favourable. However, the worry is the traditional one with politicians – what they say and what they do can be two different things.

The A.N.C. says it is strongly committed to conservation and rational use of South Africa's natural resources for the benefit of present and future generations. Its position on wildlife specifically is as follows:

'The increased pressure on land for human settlement and for agricultural production is likely to limit land availability for wildlife conservation and its use for major economic activities like tourism.... Conflicts of these land-use requirements lead to such aberrations as poaching. Yet, in many ecological zones in South Africa, wildlife management and conservation offers the only ideal balance between human economic activity through tourism and environmental conservation. Besides, wildlife is a heritage we need to preserve for posterity. There is need therefore to establish the optimal balance between devoting such lands to wildlife and meeting the requirements of human settlement and sustenance.... The A.N.C. is in agreement with the policy approach of some of the neighbouring states which advocate full community participation in management of wildlife resources and the economic benefits flowing from this resource. This approach will serve to generate much-needed income for the rural communities in the usually poor semi-arid agroecological regions like the Great Karoo, eastern Natal and around the Kruger National Park in the eastern Transvaal.'

The position papers are encouraging because, in some radical black political circles, the national parks have been looked upon as part and parcel of apartheid: in their eyes, land has been taken from the people to be used as wildlife sanctuaries for the enjoyment of the privileged and élite few.

Robinson has had direct talks with the A.N.C. and its leader Nelson Mandela on these issues.

'I am encouraged by these discussions and the position papers published by the various parties. However, I am concerned about the sincerity of all South African political parties, including the present ruling National Party Government, towards environmental policies. I do not believe environmental and conservation issues are receiving the serious thought they deserve and that all the parties to a greater or lesser degree are merely paying lip service to the immense problems we are facing. South Africa is out of step with the developed Western countries on this. Overseas, people and governments are paying sincere and serious attention to the ecological damage being done to the earth', he comments.

Robinson does not believe the Parks Board will get the R333-million loan proposed by the Board of Trade and Industries in 1990 as part of a package to boost South African tourism. Instead, he would prefer to see changes made in the National Parks Act to allow the Parks Board to borrow money to fund its developments in the open market. Robinson is confident he will be able to get US$150-million from the World Bank.

Since his appointment as Chief Executive Director he has travelled extensively overseas addressing, amongst others, the European Parliament, on future plans for conservation in South Africa. Robinson says there is an enormous amount of goodwill towards this country overseas and a determination to see that the political problems of the country are satisfactorily resolved.

He adds that the World Bank has indicated it will provide the capital the National Parks Board requires, just as soon as an interim government is installed as part of the reform process in South Africa. Having the World Bank as a major investor in the National Parks Board will also provide international support and protection for the organization should dealings with future governments turn sour.

However, the most crucial issue for the Board is to convince South Africa's black population of the need for conservation. Events in the rest of Africa have made it abundantly clear that

Every wild dog has a unique pattern on its coat and this method of identification has played a key role in the research of Africa's most endangered carnivore.

without the support of the majority of the population, the conservation effort and the national parks and game reserves will not survive. The crux of the matter lies in the concept of 'sustainable utilization' – making conservation relevant by providing material benefits from game reserves and national parks to neighbouring, impoverished communities which cast covetous eyes on the sleek game and pristine lands protected behind the high game-fences.

The issues in South Africa were extensively aired in a seminal symposium held in 1988 by the Endangered Wildlife Trust – one of South Africa's leading conservation N.G.O.s. At that conference, conservationist Ron Thomson made the point that, 'Only by adopting the attitude that wild animals are "products of the land", and not "sacred cows" to be set upon a pedestal, will many species survive this century let alone the next one. Nobody ever considers that the most utilized animals on earth – domestic stock – ever face extinction … because they don't! And the reason for this is simply because they are an integral part of man's survival kit on earth. When wildlife achieves this status in the hearts and souls of our game reserve's neighbours – be they black, white or green – it too will enjoy a similarly secure future.'

Millions of years ago, the blue layers visible in the mountain ranges of the Richtersveld National Park formed part of the ocean floor.

Thomson's viewpoint is reinforced by that of Dr John Hanks, Chief Executive of the Southern African Nature Foundation (S.A.N.F.) which is the southern African arm of the W.W.F. – World Wide Fund for Nature. The S.A.N.F. is South Africa's most important conservation N.G.O. and has raised almost R200-million in 25 years to finance more than 400 projects, many of them involving the purchase of land to be added to existing national parks and game reserves.

Hanks says the traditional approach to conservation has simply not worked, as shown by the plummeting world populations of various key species despite all efforts to save them: 'Africa south of the Sahara has 29 of the world's 36 poorest countries, where some 325 million people live in a condition of absolute poverty with annual incomes of less than US$100. As population densities build up, impoverished rural communities have no alternative but to destroy the very resources on which their survival depends, namely woodlands, grasslands and soil. With this destruction comes the inevitable loss of biological diversity, and an unjustified criticism that poor people have no love of nature', he says.

A number of official and private conservation bodies in South Africa have adopted the sustained utilization philosophy wholeheartedly. They include the Bophuthatswana, Ka-Ngwane, Gazankulu and KwaZulu nature conservation authorities, the Natal Parks Board and the Phinda Resource Reserve as well as the Conservation Corporation which runs the Londolozi and Ngala lodges next to the Kruger National Park.

Conservation Corporation Managing Director Dave Varty is probably the most vociferous and active proponent of the concept in the private sector. He has set up businesses in partnership with blacks from local communities to provide the needs of his luxury lodges such as transport and fresh vegetables. The scheme is designed in such a manner that his partners can expand their businesses to supply other customers.

One of the most encouraging stories comes from Namibia where Garth Owen-Smith and Margaret Jacobsohn have managed to reconcile the conservation needs of the desert elephant and black rhinoceros populations with the living requirements of the Himba pastoralists who inhabit this area to the considerable benefit of both parties.

However, while the National Parks Board accepts the principle of sustainable utilization, it has a different, more purist approach from that of other conservation organizations. Essentially, its view is that the Kruger and other existing parks must remain inviolate core sanctuaries and the full range of sustained utilization practices will take place in the new contractual national parks being set up. Whether this purist attitude is the best approach for South African conservation requirements is a matter of considerable debate.

Robinson says the core areas are 'sacrosanct' and will continue to be managed with as little deviation from natural ecological processes as possible. Schemes to generate extra income from hunting, and to provide material benefits to neighbours through collection of thatching grass, wood and medicinal plants, will not be allowed in the core areas but only in the new, buffer areas being added on around the fringes of the parks.

One example will illustrate the differences in approach. In the severe drought of the mid-1980s a number of the private reserves on the Kruger Park's western boundary culled game that was doomed to starve to death because of lack of grazing. The meat was sold cheaply or donated to the black communities in the area. The Kruger Park authorities did not do this but instead allowed large numbers of game, including about 10 000 impala, to die during the drought, providing food for many species of predator and scavenger which thrive during these conditions.

The reason for this was the Parks Board's philosophy of interfering as little as possible with the natural processes of its various ecosystems, and droughts are clearly part and parcel of these processes in southern Africa. The principle at stake is natural selection. The weak died in the drought, but the strong survived to pass on their characteristics for fitness to subsequent generations. Park managers were opposed to culling because it is almost impossible to discriminate between weak and strong animals; by culling randomly some genetically strong animals are removed while some genetically weak animals are able to survive to breed again.

Robinson says this approach will remain in force in the Kruger Park but what happens in the contractual parks and buffer zones on the boundaries is another matter completely. There, full sustainable utilization will be practised to provide many benefits to the local people. He rules out sport-hunting in the Kruger Park but says surplus game can be translocated to the contractual areas on the boundary and hunted there.

He also rules out such innovative money-spinners as that developed by the Bophuthatswana conservation authorities whereby hunters are brought in to dart big game that has to be translocated anyway. In the Bophuthatswana parks foreign hunters pay US$7 000 to dart a white rhinoceros – the 'kill' is accepted by international hunting organizations and everybody goes away satisfied – the hunter with his trophy, the park with the money and the rhinoceros to live happily in his new home. It may be anathema to many nature-lovers, but it is a fact that hunters are both a major source of income for the conservation cause and at the same time its natural allies – for if they do not conserve they will have nothing to hunt in the future.

Given the necessary funds, the Parks Board intends to upgrade facilities and attractions at existing parks to make them financially viable and to try to remove some of the visitor pressure from the Kruger Park. At present only three of the Board's parks – Kruger, Tsitsikamma and Golden Gate Highlands – make a profit as the rest either make losses or break even.

Robinson declines to give specific details of the Board's plans for future development of the national parks because of the infighting and ill-will this can create with the provincial and other affected conservation bodies. A major issue is how the administrative structure of the 'New South Africa' will be engineered – what will eventually happen to the four existing provincial administrations as well as to the independent and self-governing states such as Bophuthatswana, KaNgwane, KwaZulu, QwaQwa and Gazankulu, all of which have their own nature conservation organizations?

It seems likely these independent and self-governing states, which were products of the 'grand' apartheid scheme of providing each tribe with its own homeland, will be reincorporated into South Africa. That being the case there are a number of obvious targets for the National Parks Board. Both Gazankulu and KaNgwane have proclaimed game reserves on the boundaries of the Kruger Park making them prime candidates for the kind of buffer-zone development Robinson has in mind. QwaQwa has proclaimed a 20 000-hectare nature reserve bordering the Golden Gate Highlands National Park and the Parks Board is already providing advice on how to manage it. It is clearly another natural candidate for inclusion.

What happens with the four provinces is considerably less certain. Natal, for example, is in the anomalous position of having no national parks at all, despite that province's superb Drakensberg scenery and the astonishingly rich ecosystems in northern Natal and KwaZulu. Instead, most of the game and nature reserves here are run by the Natal Parks Board which, until Robinson's appointment, had been far more innovative and flexible in their approach than the National Parks Board. Robinson has great respect for the Natal Parks Board and the way it does things and feels that the organization's independence should be maintained.

In addition to the provincial conservation bodies, the independent and self-governing state conservation organizations and the National Parks Board, the various local government authorities known as regional services councils also have their own conservation departments. It is a complex situation which arouses petty jealousies and invites dissension and infighting. It is a far cry from the ideal of having everybody working together for the common good of the conservation cause.

Bringing about some kind of rational order for conservation in South Africa is a task that has been evaded time and again by the Government, most recently in the 1991 President's Council report on the environment. Whether the issue will be tackled properly when a new democratic order finally prevails in South Africa remains to be seen.

While one can argue over the specifics of the National Parks Board's plans for sustained utilization it is extremely encouraging that the organization has seen, and accepted, the need for change and is rapidly doing something about it. Other positive developments include the growing involvement of the private sector in the conservation effort through groups like Gold Fields of South Africa funding education centres throughout the country, Rand Mines contributing land and the S.A.N.F. aggressively funding the purchase of land. On their shoulders lies the burden of ensuring that the natural wonders and enormously rich biodiversity of South Africa remain with us and the generations to come.

WEST COAST NATIONAL PARK

The development of the Langebaan Lagoon and surrounding areas into the West Coast National Park makes a classic case study for environmental managers because it involves the economic, ecological, bureaucratic and sociological aspects of the struggle to conserve threatened natural areas.

That it is a success story is thanks largely to the efforts of the National Parks Board and the Southern African Nature Foundation (S.A.N.F.) which latter organization stepped in to provide the funds to buy a number of the key farms bordering the lagoon. This was necessary because Government finance was not available. However, the story has not yet reached its conclusion and this will come only through a change in the attitude of the majority of people living in the south-west Cape to their natural environment.

West Coast National Park warden Sarel Yssel sums up the prevailing attitude as one of overwhelming self-interest and lack of consideration: 'There seems to me to be an alarming lack of knowledge and understanding of conservation in the marine environment, and this is coupled with a very selfish attitude prevalent amongst people such as fishermen who use the sea's resources. Part of the problem here is that the West Coast Park is a young national park which has been in existence only since 1985. We are imposing regulations in an area where people previously did more or less what they felt like. They do not see the valid conservation reasons for the regulations and feel that they are there to be broken. That attitude has to change.'

It is an attitude that has prevailed through much of the history of the Langebaan area, as when hundreds of penguins and cormorants were needlessly killed during the blasting that took place in the mid-1970s to create a deep-water channel into Saldanha Bay for huge bulk carriers. The explosions killed vast

A bank cormorant which is found only along the west coast of South Africa and Namibia, displays from its nest on one of the West Coast National Park's offshore breeding islands.

numbers of fish which attracted the penguins and cormorants to the site of the blast just in time to be killed by the next explosion. The solution was twofold and simple: ornithologists were able to suggest blasting times when the penguins and cormorants were less likely to be feeding and sufficient time was allowed between blasts to permit the birds to return to shore; tape-recordings of killer-whale calls were also played under water and frightened off the birds when blasting was imminent. Conservationists were left wondering why their advice was not sought in the first place, however.

Then there was the inexcusable six-year delay in building a predator-proof wall on the causeway that was constructed in 1976 to link the mainland to Marcus Island to provide a breakwater for the harbour at Saldanha. The causeway allowed predators such as grey mongooses, genets and Cape foxes on to the island where they wreaked havoc among the nesting sea-birds. The birds bred there since it was, to them, a traditional island breeding-site safe from predators. The need for the wall was foreseen by ornithologists but nothing was done about it by the developers. Populations of breeding sea-birds on Marcus Island are only now recovering and still all the predators have not been eliminated.

Growing interest in the West Coast by property speculators and developers and the fact that Langebaan is barely an hour's drive from Cape Town mean that the human recreational pressure on the resources of the national park must increase. And that is why the Parks Board has made environmental education a priority for the West Coast National Park. It uses the education centre at the restored Geelbek farmhouse which was funded by the mining house Gold Fields of South Africa.

The facts of daily life in the park reflect in microcosm the effects of the greater developments taking place around it. These include the enormous increase in property development and speculation along the West Coast and the overexploitation of fish resources in the rich fishing grounds offshore. The property boom, by dramatically driving up land values, has made it

LEFT: *Following winter rains in the western Cape, spring wildflowers coat the Postberg section of the West Coast National Park.*

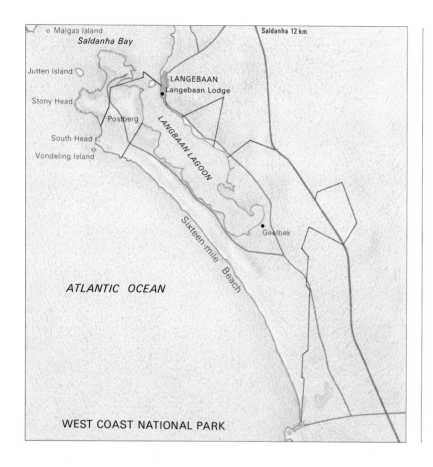

Malgas Island
Saldanha Bay
Saldanha 12 km
Jutten Island
LANGEBAAN
Langebaan Lodge
Stony Head
Postberg
LANGEBAAN LAGOON
South Head
Vondeling Island
Geelbek
Sixteen-mile Beach
ATLANTIC OCEAN
WEST COAST NATIONAL PARK

RIGHT: *Kelp gulls soar over Langebaan Lagoon. A major breeding colony of these birds is found on Schaapen Island.*

virtually impossible to achieve the initial objectives for the West Coast National Park which were to extend it down the coast past the village of Yzerfontein to Bokpunt.

The extent of marine overfishing is immediately apparent to researchers in the park who are monitoring the breeding success of birds such as the Cape, bank and crowned cormorants and the jackass penguin. These birds prey on the same anchovy and pilchard species sought after by the trawlers. Dramatic falls in the breeding success of these birds have been caused by over-fishing because the birds have been unable to find enough food to sustain them during the months they are forced to remain in close proximity to the breeding islands.

A number of areas of State land were earmarked in 1980 for inclusion in the national park when its development was accepted in principle by the Government; however these were subsequently not included. The Donkergat section at the tip of the Langebaan Peninsula was allocated to the South African Defence Force while the Department of Sea Fisheries retained control of two of the offshore islands, Dassen and Vondeling, that were also supposed to be included in the national park.

Low tide exposes vast stretches of salt marshes which are essential for a healthy ecosystem in the Langebaan Lagoon.

Salt marshes dominate the southern end of Langebaan Lagoon in this aerial view looking north towards Saldanha Bay.

The nature area was proclaimed in 1984 and was followed in August 1985 by the proclamation of the national park. At that time the park included the lagoon up to the high-water mark, Malgas, Jutten, Schaapen and Marcus islands, and a strip of beach, the Admiralty zone, along Sixteen-mile Beach. The position was far from ideal in that the Parks Board did not own or control the ground surrounding the lagoon. This meant that it could not manage the ecosystem as it would like to do because it had no control over land access to the region. In addition, the Government made it clear it had no funds to buy further land.

At this stage, however, light began to appear at the end of the tunnel and the essential core of the park was secured through a combination of money from private enterprise and innovation by the Parks Board in negotiating the first South African instance of inclusion of privately owned ground into a national park on a 'contractual' basis.

The Southern African Nature Foundation, which is the local arm of the World Wide Fund for Nature (W.W.F.), identified the West Coast National Park's land requirements as a priority and acquired the farms Bottelary, Geelbek, Abrahamskraal, Zeeberg and Schrywershoek for the park. That generous act secured the land around the southern end of the lagoon where the great majority of the sensitive salt marshes are found.

The Parks Board drew up the contractual agreement with the Oudepost syndicate, the owners of the 1 800-hectare Postberg

Nature Reserve on the Langebaan Peninsula which forms Langebaan's western arm jutting into Saldanha Bay. In terms of the agreement the syndicate members, who are mainly wine-farmers, retain private title to their land which is now managed by the Parks Board. An embargo has been placed on the construction of new buildings in the reserve which is now open to the public for the months of August and September each year, for the duration of the spectacular wild-flower season.

Despite the problems, what has been achieved is the safe-guarding of one of South Africa's most important coastal wetlands – one which is crucial to the survival of a number of endemic sea-birds and also as a wintering ground for tens of thousands of waders which migrate here to escape the bitter Northern Hemisphere winters. The Langebaan Lagoon shelters around 60 000 of these waders and, because of this, has been registered under the Ramsar Convention to which South Africa is a signatory. This Convention aims to identify 'Wetlands of International Importance, Especially as Waterfowl Habitat', with a view to ensuring their adequate conservation. Langebaan supports more birdlife than any other wetland in South Africa.

Salt marshes are rare around the long but rugged South African coastline and Langebaan Lagoon contains by far the largest tracts. The lagoon is unique in South African terms in that it is not an estuary – in other words it is not fed by a river. It was formed when rising ocean levels allowed the seas to break

Jackass penguins scramble into the cold waters of the Atlantic Ocean. The penguins forage underwater and return to land to roost and breed.

through the granite headlands that flank the entrance to Saldanha Bay and flooded the area lying behind the coastal dunes that today make up the Langebaan Peninsula.

The lack of fresh water flowing into the lagoon has naturally resulted in relatively constant salinity conditions, while variations in water depth in the lagoon are more stable being controlled solely by the tidal fluctuations and not influenced by river flooding. These factors have allowed full development of the various plant communities that make up a salt marsh. The lagoon teems with life; to give but one example, more than 550 species of invertebrates have been identified here, nearly double the number found in any other South African lagoon.

The vegetation on the land surrounding the lagoon is known as West Coast Strandveld; this is a major veld-type subdivision of the Cape Floral Kingdom, generally regarded as one of the world's richest areas of plantlife. The western Cape has more than 8 500 species of plants in an area just 90 000 square kilometres in extent; around 5 850 of those plants are endemic, being found nowhere else in the world. The area explodes into colour during the western Cape spring months of August and September which accounts for the popularity of the Postberg reserve.

However, the main attraction of the West Coast National Park for most visitors lies in its birdlife in the form of the migrant waders and the sea-birds that breed on the islands and feed on the fish wealth of the neighbouring ocean.

Jutten, Schaapen, Malgas and Marcus islands are home to approximately 50 per cent of the breeding population of the local subspecies of swift tern, 35 per cent of the world's population of Hartlaub's gull, 12 per cent of the world's black oystercatchers, and 10 per cent of the world's Cape, bank and crowned cormorants. It also has 25 per cent of the world's Cape gannet population in the form of the 60 000 birds that breed annually on Malgas Island.

Also breeding here are colonies of the jackass penguin which is listed as 'vulnerable' in the *South African Red Data Book – Birds*. Its total population has crashed from perhaps several millions at the turn of the century to around 120 000 at present and unfortunately numbers are still falling. The reasons for the initial crash were a combination of large-scale egg-collecting and the exploitation of the guano deposits for fertilizer which removed the thick bed of dried bird manure in which the penguins dug their nesting burrows.

Egg-collecting is now illegal and guano-scraping has been stopped on the breeding islands, but penguin numbers continue to drop; this is probably because the birds cannot find enough food because of overexploitation of the fish stocks by commercial trawlers. Being unable to fly, the penguins are more dependent on fish shoals found close to the breeding islands than are the gannets and cormorants which can range further out. Oil spills are another important cause of penguin mortality.

Burchell's zebra in the Postberg section of the park. It is distinguished from the mountain zebra by the greyish stripes which occur between the dark stripes on the hindquarters.

A black oystercatcher broods its eggs which are superbly camouflaged amongst the surrounding shingle and broken shells when the bird is off the nest.

Ringing of individual birds has shown that the bulk of Langebaan Lagoon's waders come from the Siberian tundra area of northern Russia. The birds spend between September and April at Langebaan, fly the 7 500 kilometres to Siberia in May and early June, breed there during the months of June and July and fly back to Langebaan during August and early September.

According to Professor Les Underhill of the University of Cape Town, more than 30 000 waders have been ringed over the past 20 years by the Western Cape Wader Study Group and the hundred or so recoveries in the Northern Hemisphere have given valuable information about the birds' habits. Underhill points out that curlew sandpipers, the commonest migrant wader at Langebaan, fly the direct route up Africa along the Rift Valley then across the Middle East and through Central Russia to the extreme north of Siberia. Knots, however, go westwards around the coast of West Africa and come in across Western Europe while sanderlings go up the west coast to Nigeria and then fly straight across the Sahara into Western Europe.

The waders eat around 150 tons annually of the tiny animals that thrive in the lagoon such as the minute snail *Assiminea* but return around 44 tons of fertilizer to the salt marshes in the form of guano. That in turn fertilizes and supports the salt marshes on which the birds depend.

Through intense feeding the waders put on large fat reserves before leaving Langebaan in the autumn and a curlew sandpiper just before departure will weigh up to 100 grams, about double its early summer weight. Underhill calculates that this

load of 'fuel' takes them about a third of the way, and that somewhere around the equator they have to spend about two weeks feeding up again before continuing their journey. Another feeding session is required somewhere before the birds take off on the final leg to the breeding grounds. The survival of these birds depends on an international effort to protect not only their breeding grounds in Siberia and their wintering grounds in the Southern Hemisphere, but also the wetlands that provide the 'refuelling' stops in between.

The Parks Board has tried to accommodate human activities in the lagoon as far as possible. The lagoon is zoned into three sections, with Zone A at the head of the lagoon designated a multipurpose recreational area where activities such as motorboating, water-skiing, fishing and diving are allowed. Zone B is exclusively for sailing boats and windsurfing, while Zone C at the foot of the lagoon is a wilderness area which no-one may enter. The Parks Board has allowed traditional net-fishing by permit-holders to continue and allows anglers to take bait from the lagoon in limited quantities.

Ironically, birdwatchers are the interest group probably the least catered for in the West Coast National Park because of necessary restrictions on access to important breeding and feeding areas. This, however, is soon to change. The Parks Board plans to restore the old farmhouse at Bottelary as a lodge for birdwatchers, and to put in a special 'bird walk' and hide at Geelbek. There are also plans to construct a series of bird hides along the salt marshes.

Access to the bird islands will continue to be restricted, for a very good reason. Disturbance by people walking too close to cormorant colonies results in the birds leaving their nests and, within seconds, hovering kelp gulls swoop in to steal the eggs. Schaapen Island has the largest breeding colony of the southern African subspecies of the kelp gull and their numbers are such that continued disturbance of the cormorant and penguin colonies would cause substantial damage to these species.

The Cape gannet, however, is totally unperturbed by man's presence and reacts to a close approach with a swift and painful jab of the beak at the nearest leg. No kelp gull in its right mind tries stealing eggs from these colonies and it is possible that in due course small groups of birdwatchers under Parks Board supervision may be allowed to visit Malgas Island.

The spectacle that awaits them of thousands upon thousands of breeding gannets underlines the wealth of this ecosystem. The colony of closely packed birds, each nest built just outside lunging and pecking range from its neighbours, stretches in a white mat in all directions. The smell from the guano is overpowering and there is a constant deafening roar of sound. A steady stream of returning birds passes overhead before plunging into the colony in the gannet's version of a perfect three-point landing while at the 'runway' other birds line up for take-off and the search for more food.

Visitors will see the effects of marine pollution along our coastline, for here and there in the colony stand gannets with their natural gleaming white plumage stained brown – the result of diving into oil slicks while chasing fish. The same thing is happening to penguins and cormorants that fish offshore. Some succeed in returning to land where, if found in time, they are taken to a facility in Cape Town run by the South African National Foundation for the Conservation of Coastal Birds (S.A.N.C.C.O.B.) to be cleaned up and returned to the wild.

Hanging over the West Coast National Park is the constant threat of an oil spill from the bulk carriers using Saldanha Bay. Should a large spill ever occur, there is virtually nothing the park authorities could do to prevent the oil spreading into Langebaan Lagoon because the power of the tide surging in and out of the mouth of the lagoon makes it impossible to keep oil-restraining booms in place during clean-up operations.

Such a threat, and the sickening sights of oiled, choking and dying sea-birds, should reinforce the determination of all concerned about the environment to do what they can to fight pollution and protect this superb national park.

Thousands of gannets form a raucous breeding colony on Malgas Island.

TANKWA KAROO NATIONAL PARK

Although this interesting national park was proclaimed in 1986, it will take some time before the general public will have access to the area as the Parks Board has decided that rehabilitation of the veld and the acquisition of further land is necessary before the development of a tourist-orientated infrastructure is justified.

The 27 000 hectare Tankwa Karoo National Park is situated about 250 kilometres north-east of Cape Town, and 90 kilometres south of the small town of Calvinia. It lies on the western boundary of the Great Karoo in a veld-type known as Succulent Karoo or winter-rainfall Karoo because it falls within the winter-rainfall belt of the western Cape.

Succulent plants survive in the Karoo by storing water to protect themselves during the drought periods.

Until the proclamation of the Tankwa Karoo Park, this was a veld-type sorely in need of protection. Although it covers about 3,7 million hectares of South Africa's land surface area (about three per cent of the total), the only formally protected area of Succulent Karoo was a tiny 200-hectare patch in the Gamka Mountain Nature Reserve. This is one of the main reasons the Parks Board agreed to take over the Tankwa Karoo land as a national park after the Cape Provincial nature conservation authorities refused it on the grounds that its condition was so poor that it would require a disproportionate amount of already scarce financial and manpower resources to rehabilitate it.

It has to be admitted that the Tankwa Karoo National Park as it now stands seems a poor choice for a national park because, while the Succulent Karoo veld-type as a whole contains an abundance of endemic plant species requiring protection, not many of them appear to be found in this park. Further, the diversity of plant species at first glance does not appear to be at all impressive although intensive plant studies have not yet been carried out and the area has been very seriously overgrazed by sheep and goats.

LEFT: *Under a sky teasing with the promise of life-giving rain, parched clumps of yellow grass stretch to the horizon in the Tankwa Karoo.*

Parks Board Chief Executive Director Dr Robbie Robinson concedes the point but indicates that the Board has long-term strategic plans for this area, and it could take another 30 to 50 years before the development of this national park is complete. 'The Tankwa Karoo Park covers an area that deserves to be protected but, as it now stands, I cannot see that it really meets the criteria that I look for in areas deserving national park status', comments Robinson.

However, he has his sights set on the entire Tankwa basin which covers some 200 000 hectares locked between the Cederberg Range to the west, the Roggeveld Mountains to the east and Swartruggens Mountains in the south-west. He feels that the basin is unique and, if conserved as a unit, would create a wilderness area in which the atmosphere of desolation and remoteness would complement the preservation of a semi-desert ecosystem with the particular attraction of the wild flowers that bloom in spring after the winter rains have fallen.

He points out that, like the Richtersveld, the human carrying capacity of this region will always be very low, and thus sophisticated facilities will never need to be developed to cater for large numbers of visitors. Instead, the tourists attracted to this area will be those who can appreciate the remoteness and silence of a totally different kind of environment as well as those who are primarily interested in the plants.

Acquiring the rest of this ground from the local small-stock farming community might appear a tall order but Robinson seems confident that the changes taking place in South Africa, with the Government forced to re-think its budgetary priorities, will play into his hands. He believes, for example, that the Government is going to have to cut farming subsidies in marginal areas. 'The South African agricultural industry is fraught with government subsidies with most of them going to farmers operating on marginal land that should not have been farmed in the first place. This has been a very irresponsible policy but it seems it is now changing – something that should have been done long ago. When the subsidies eventually fall away, mar-

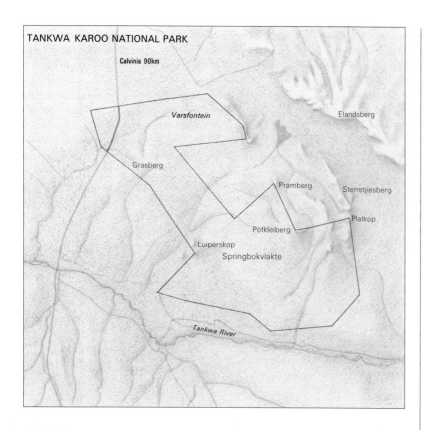

TANKWA KAROO NATIONAL PARK

Calvinia 90km

Varsfontein

Elandsberg

Grasberg

Pramberg

Sterretjiesberg

Platkop

Potkleiberg

Luiperskop

Springbokvlakte

Tankwa River

ginal farming areas will no longer be paying propositions. This in turn means these farms should become available for other uses such as conservation,' he says.

If this grand plan eventually falls into place then the large desert-adapted game species that used to live here, gemsbok and springbok for example, will be reintroduced. In the meantime, the only mammals to be seen are those which have managed to adapt to man's use of the land, such as spring-hare, baboon, grey duiker, aardwolf, bat-eared fox, porcupine and mongoose, or which live in areas which cannot be cultivated, such as klipspringer and rock dassie. Then there are those of course which have managed to survive in the Karoo despite every man's hand being raised against them – the caracal, black-backed jackal, and Cape fox.

What the Tankwa Karoo certainly has in large measure is a feeling of remoteness as well as utter silence. Driving in from Cape Town the tourist works his way through the spectacular scenery of the Cape Fold Mountains and the lush wine estates until he passes the town of Ceres and gets north of the Hex River Mountains. Then abruptly, in the rain shadow of the mountain ranges, the Karoo starts and beyond Hottentotskloof at Karoopoort the tar road degenerates to a gravel road that runs

The distinctively tassel-eared caracal is another of Nature's born survivors, holding its own in farming areas despite every man's hand being turned against it.

The rock hyrax or dassie thrives even in heavily cultivated areas because its rocky retreat provides protection from agricultural development.

The Cape fox is one of the most attractive of the small carnivores but is seldom seen outside of the national parks because persecution has taught it extreme caution.

virtually arrow-straight across the Tankwa Karoo heading for Calvinia, 220 kilometres to the north. There is nothing here but the odd farmstead – often deserted – and the more frequent windmills rising forlornly above the flat landscape with the tanks that they feed with life-giving water standing nearby.

Signs of life are few and far between – the occasional flock of sheep and every now and again a pale chanting goshawk standing on its distinctive long red legs on the top of a telephone pole alongside the road. At the end of summer, before the winter rains come, the picture is one of complete desolation as dust devils twist through the heat haze across the flat plains where the only vegetation consists of Karoo bushes and the occasional clump of bright yellow bushman grass.

The average annual rainfall for this region is a miserly 50 millimetres and even this is of erratic occurrence. In the good years the succulent plants burst into brief flower, spreading a gay face over the land before it reverts to drabness again and the long wait for the next rains. The Tankwa River is usually dry and lies outside the present boundaries of the park.

The pale chanting goshawk does most of its hunting from an exposed perch from which it will glide slowly down when prey, usually a lizard or small mammal, is spotted.

*When it does rain in the Tankwa Karoo, the result is a
colourful but short-lived display from plants such as this
mesembryanthemum species.*
RIGHT: *The Karoo as it is for much of the year – a dry, empty
land stretching as far as the eye can see.*

Astonishingly, however, in the midst of this arid landscape one
can occasionally hear frogs croaking because of the existence of
six natural springs or fountains. These provide enough water to
maintain at least a few square metres of bright green reed-beds
even in the worst drought.

Since acquiring the Tankwa Karoo Park the only development undertaken by the Parks Board has been the upgrading of
the boundary fences to make sure all are jackal-proof following
complaints from neighbouring stock-farmers. A caretaker
ranger looks after the area which will lie dormant, slowly recovering from the overgrazing of the past, until the Parks
Board's grand scheme for the region is brought to fruition.

RICHTERSVELD NATIONAL PARK

The proclamation of the Richtersveld as South Africa's first wholly contractual national park has been widely acclaimed as the National Parks Board's breakthrough into conservation in the 'New South Africa'. However, it took 18 years of negotiation as well as legal action by the community of some 2 000 Nama pastoralists and farmers before the Parks Board metaphorically threw its book of rules out of the window and reached a compromise in July 1991. Under the agreement, the Richtersveld became, with certain modifications, a 'Schedule 5 national park'.

To show just how different conservation in the Richtersveld will be compared with a 'traditional' Schedule 1 national park, such as the Kruger Park which is wholly controlled by the Parks Board, consider that in the Richtersveld the Nama pastoralists will be allowed to graze their goats, sheep and cattle in the park and to bring in their dogs to protect and herd the flocks. Rogue predators such as caracal and black-backed jackal will be trapped and removed by Parks Board officers if these attack their flocks. Other traditional rights such as that of gathering firewood will also be permitted. The Richtersveld will be run by a Management Planning Committee made up of four National Parks Board staff and five Namas representing the Richtersveld community.

Displeasing though it may be to conservation purists, mining will be a fact of life in the new national park which park managers will have to live with. Park authorities, however, will have some say over proposed new developments and the rehabilitation of worked-out areas. Two diamond-mining companies are continuing already established operations along the banks of the Orange River within the Richtersveld and other companies are prospecting the region for base metals. They might well find something because small – although econom-

The unique flowers of Hoodia bainsii, *a leafless succulent which occurs mainly in the dry regions of South Africa.*

ically insignificant – quantities of minerals have already been discovered in the Richtersveld; there is also an active lead and zinc mine, Rosh Pinah, which is found just north of the Orange River inside Namibia.

The reason that the Parks Board has accepted these conditions is that, despite all the drawbacks, the Richtersveld is still so environmentally rich that it deserves the highest level of protection the country can offer. Its semi-desert scenery matches the best of the rocky deserts in neighbouring Namibia while it nurtures an extraordinary wealth of plantlife, much of which is endemic to the Richtersveld.

The unique Richtersveld landscape is characterized by the combination of its many-hued, rugged mountain ranges and the distinctive outlines of a number of its larger plants such as the quiver-tree or 'kokerboom' (*Aloe dichotoma*), the elephant's trunk or 'halfmens' (*Pachypodium namaquanum*) – Afrikaans meaning half-person – and the bastard quiver-tree (*Aloe pillansii*). Seen in the quality of light that is typical of desert afternoons and early mornings, its beauty is such that the visitor to this area could almost forget that he is in a harsh wilderness ... which would be foolish.

Summer temperatures often soar as high as 40 degrees Celsius and there is little shade to be found: there are no trees other than the odd large quiver-tree and even they are few and far between. To become lost or suffer a vehicle breakdown without having enough water available could prove fatal in this pitiless environment. Dust storms are common and when strong winds blow they can propel the grit with sufficient force to sandblast the paint off exposed vehicles.

Sleeping on the ground here is not recommended, unless in a tent with a built-in groundsheet, because the Richtersveld is home to several of the world's most poisonous scorpions. Various members of the family Buthidae, the thick-tailed scorpions, are extremely common. The most poisonous are members of the genus *Parabuthus* and some of them have the ability to squirt their venom up to a distance of about a metre.

LEFT: *The dry De Hoop plains stretch towards towering mountain ranges which have a wetter microclimate created by condensation from fog coming in off the cold Atlantic ocean.*

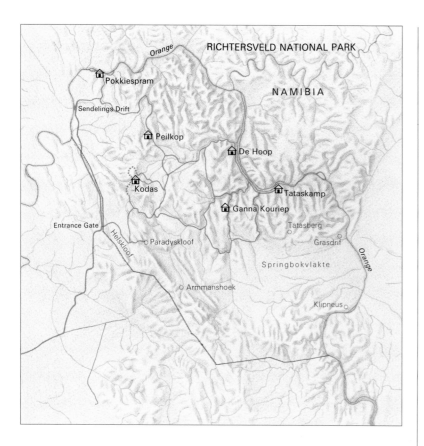

RICHTERSVELD NATIONAL PARK

NAMIBIA

That the Richtersveld has become a national park is due in large part to the sensitivity, patience and foresight of present Parks Board Chief Executive Director Dr Robbie Robinson who, in his previous post as head of coastal parks, was able to break through the stalemate and antagonism that existed in 1989 between the Parks Board and the Nama residents.

The Parks Board had originally planned to acquire the Richtersveld as a conventional Schedule 1 national park from which the pastoralists would be removed. Alternative land was offered to them in the form of the 'Corridor West' farms closer to the coast – 70 000 hectares of State land inland of Port Nolloth and Alexander Bay. But most of the pastoralists rejected this proposal. When they first heard that the Parks Board intended removing them from their land in the interests of conservation, one pastoralist is reported as commenting: 'Hulle gee om vir die halfmens, maar wat van die volmens?' – 'They care about the half-person but what about the whole-person?'

Many of the residents were not, in fact, opposed in principle to the concept of a national park in the region but they did want to retain access to it for their livestock. Paulus de Wet, a farmer from the tiny village of Khubus and now a member of the Management Planning Committee, commented during negotiations in 1990: 'I am not opposed to the formation of the park but I still want to farm in the specific area. I want to maintain my freedom and access that I have as a farmer to move to where

there is grazing in the veld. Why can the conservation of plants and the farming of stock not occur at the same time?'

In its early negotiations the National Parks Board delegates dealt exclusively with the local-government structure, specifically the management board for the northern Richtersveld, and not with the community as a whole. All seemed to be in order and the Parks Board had received Government permission to go ahead with removing the people when, in early 1989, the community applied for an interdict to stop the contract from being signed. That led within a month to direct negotiations with the community by the Parks Board.

The agreement finally brought about set up the Richtersveld National Park for an initial period of effectively 30 years – 24 years together with a six-year notice-of-termination period – with the running of the park to be decided by the Management Planning Committee. The community retains the right to cancel the agreement after 24 years, subject to the notice period, if they are unhappy with the way things have worked out.

The Parks Board will pay an initial annual fee of 50 cents a hectare – about R80 000 in total – for the right to use the land as a national park; the fee is subject to review and will be linked to the rate of inflation. The money will be paid to the Richtersveld Community Trust which has been set up to employ it in the best interests of the local Nama community. In a land where the average income is R90 a month, this represents a substantial sum of money. In addition, the Parks Board has agreed to set up a nursery growing the indigenous succulent plants of the area for sale to visitors; the profits from this will also go to the Richtersveld Community Trust.

The Nama pastoralists also retained the right to graze their stock in the area but maximum limits were set at 6 600 small stock units. A small stock unit is a goat, and a cow or donkey is the equivalent of seven small stock units. Should any pastoralist voluntarily leave the area to go to Corridor West, these maximum levels will be reduced accordingly, and permanently.

Under the agreement, the Nama members of the Management Planning Committee report back to the community, and the herdsmen are bound to co-operate with the management decisions reached by the Committee. The management plan will eventually identify core wilderness areas from which the flocks must be excluded. Other areas in the park will also be fenced off to protect some of the extremely rare plants which, in some cases, are restricted to just two or three colonies.

It is possible, although far from certain, that the number of pastoralists in the Richtersveld will drop in years to come because the younger people may prefer to find other, easier ways of making a living and move to towns. However, some farmers have indicated they may yet move to the Corridor West ground that remains on offer and thus the pastoralist numbers within

The Richtersveld is a land of ever-changing light, particularly at sunset when the multi-hued sky also means relief from the day's scorching heat.

the park may eventually decline without any suggestion of sociologically or politically unacceptable forced removals.

In addition to the money flowing to the Richtersveld Community Trust, the Nama residents are to be accorded preference in the job opportunities which will be created by the park. When fully developed it is anticipated that the park will have a staff complement of around 40. Their land will receive protection from the illicit plant-collectors and four-wheel-drive enthusiasts whose uncontrolled activities were formerly among the major threats to the region.

The Parks Board's development plans are tailored to keep intact the Richtersveld's aura of desert wilderness and solitude. There are of course some very real dangers to visitors and consequently they will not be allowed to wander alone through the region. Instead, organized walking trails will be instituted, where hikers will be accompanied by Parks Board guides.

A number of hikes will be laid out, operating from two base-camps. There will also be two larger rest-camps but each will consist of only five huts providing accommodation for 10 people. Facilities will be kept simple and basic, with the design

for the huts for the rest-camps and hiking base-camps copying the 'matjiehuis' used by the Nama farmers. This is a beehive-shaped hut made from a wooden frame which is covered by reed mats and skins. The matjiehuis is surprisingly cool in the scorching Richtersveld summers because the reed construction allows whatever breeze is present to blow through the hut.

It is intended that permits will be issued to allow fishing in the Orange River which flows in a 100-kilometre loop north around the Richtersveld. The river contains some monster specimens of sharp-tooth catfish ('barbel') weighing up to 25 kilograms, as well as large-mouth yellowfish of 10 to 12 kilograms. The Orange River will also be opened to canoe safaris run both by the National Parks Board and by recognized private operators who will have to adhere to a code of conduct.

The main attractions of the area for the tourist are its scenery and plantlife. There is no big game spectacle because the Richtersveld's carrying capacity is naturally low. In addition, the reintroduction of species known to have been historically present, together with an expansion in numbers of those which have never been eradicated from the park, will depend on the impact

Eternal scavenger and survivor in the most inhospitable of regions – the black-backed jackal.
LEFT: *Barren country similar to the Richtersveld continues north of the Orange River, extending into Namibia and eventually merging into the Namib desert.*

of the pastoralist farmers. If, after allowing for their activities, ecologists judge there is not enough grazing to spare, then additional game will not be introduced. The conservationist's priority in the Richtersveld is to restore the veld and repair as fast as possible the damage caused by previous overgrazing. Game species that could be reintroduced into the park include gemsbok, springbok, eland and Hartmann's mountain zebra.

One of the most important factors in the development of the rich plantlife of the Richtersveld is the existence of localized microhabitats. These have been created by the rugged topography of the region which causes quite dramatic climatic dif-

ferences over relatively short distances. The mountain ranges forming the backbone of the Richtersveld rise to about 1 000 metres above sea level and the average annual rainfall on their peaks can be as high as 300 millimetres compared with as little as 15 to 30 millimetres in the lower-lying areas along the valley of the Orange River. Although the Richtersveld falls within the winter-rainfall region of South Africa, it receives only a fraction of the precipitation falling in the south-west Cape. This is caused by the cold Benguela Current running northwards along the coast which inhibits cloud formation and thus limits the amount of rain that falls inland. However, the cold moist air coming in from the sea frequently forms mists and fogs which condense on the mountain tops and sustain many of the peculiar succulent plant species that have evolved here.

Most of these plants belong to the family Mesembryanthemaceae, commonly known as mesems. This is the world's largest family of succulent plants and the majority of its species are, with few exceptions, endemic to South Africa. Many of them are extremely colourful, bursting into flower during the rainy season from May to September in order to attract the bees, flies, wasps and moths that pollinate them and ensure their continued survival. Amongst this family are the various species of stone-plants (*Lithops* spp.), amazing little plants which rely on camouflage to avoid being eaten by herbivores. The fleshy leaves are almost identical with the pebbles surrounding the plant which is thus rendered invisible to the eye of the untrained observer.

A number of the larger plants in the Richtersveld, such as the halfmens and the quiver-tree, can survive years of scorching heat with virtually no rainfall. Judging by the growth of measured specimen plants it is possible that the halfmens takes hundreds of years to reach its maximum height of almost three metres. The name comes from a Nama superstition that these plants are transformed people.

The bastard quiver-tree (*Aloe pillansii*) is much rarer than its better-known relative, the ordinary quiver-tree (*Aloe dichotoma*), which is found throughout much of the north-western Cape and southern Namibia. The former can grow to a height of 12 metres and can readily be distinguished from the quiver-tree in silhouette because its crown has fewer and more erect branches.

Another very distinctive aloe is the reddish-coloured *Aloe pearsonii*. This yellow- or red-flowered species clothes a number of hillsides near Helskloof through which runs one of the access roads into the interior of the park. In the same area the visitor will also see the maiden's quiver (*Aloe ramosissima*) which has

the same distinctively shaped leaves as its larger relatives but forms a broad, low bush instead of growing into a tree.

Botanists are still discovering new plants in this area with examples over the past decade including *Aloe meyeri*, *Tylecodon kritzingeri* and *Portulacaria armiana*. The area is a botanist's paradise but a worrying aspect is the alleged dearth of young plants in the Richtersveld. It is said, for example, that there has been no germination of seedling bastard quiver-trees for the past 30 years. This is certainly not true as young plants of this species are simply difficult to recognize when small, but nevertheless there is a strong likelihood that recruitment of many plants has been adversely affected by overgrazing. Goats in particular often selectively eat flower-heads and thus reduce seed production. If these fears are well founded, the declaration of this national park has come not a moment too soon.

LEFT: *According to local legend, the halfmens is half man, half tree. An unusual feature is that the tip always inclines to the north.*

The swallow tailed bee-eater nests in burrows made in sand banks. As its name suggests, its diet consists of bees and wasps and other insects.

KALAHARI GEMSBOK NATIONAL PARK

Welcome to another world ... a world that is vastly different from the normal picture painted for overseas tourists of a lush sub-tropical South Africa, of a paradise basking in sunshine, teeming with game and fringed by golden beaches and dramatic coastlines.

The Kalahari of the northern Cape Province and Botswana is a harsh, arid, remote, dusty region; it is a place where visitors can expect to bake during the day and freeze at night. It also happens to be a unique and fascinating ecosystem, one of the world's few wilderness areas to have survived largely intact into the last decade of the 20th century. The name is derived from the word 'Kgalagadi' whose meaning is unknown, but which is the name of those parts of central Botswana inhabited by the Bakgalagadi people. It has an almost magical ring to it and visitors prepared to look beyond the initial, hostile surface impressions will find the Kalahari a magical and beautiful land. Often described as a 'desert', it is perhaps better categorized as arid savanna and it possesses a rich and varied wildlife superbly adapted to survive its implacable environmental extremes.

There is no better way to experience the Kalahari than by visiting the Kalahari Gemsbok National Park. The essential attraction of the park is that it offers the ordinary tourist ready and safe access to this desert wilderness without the need to mount a full-scale expedition. Visitors can fly into Upington in the northern Cape, hire a car and be in the park four hours later. The park's infrastructure provides accommodation in three rest-camps, as well as a restaurant and shops stocked with essential requirements.

The alternative would be to visit one of Botswana's Kalahari reserves, but for these a four-wheel-drive vehicle is required and the intending visitor will have to provide all living necessities. The huge Gemsbok National Park and Central Kalahari Game Reserve are closed to the general public and there is no accommodation or other infrastructure in the other reserves. Once in, the visitor is on his own. The decision to go alone into game reserves like Khutse or Mabuasehube is not one to be taken lightly in this hostile environment and most visitors travel in groups to provide the security of a back-up vehicle.

Pearlspotted owls are often seen roosting in trees during the day at Nossob Rest-camp.

After all that preparation, one would normally expect to see less wildlife than in South Africa's Kalahari Gemsbok National Park, probably because in the latter the Nossob and Auob river-beds tend to concentrate the antelope and other creatures during the rainy season. The three rest-camps at Nossob, Mata Mata and Twee Rivieren are all more than a hundred kilometres from each other, constituting oases from which the visitor can explore this wild and beautiful land.

And what an experience that can be. The Kalahari appears barren at first glance and yet, if you are in the right place at the right time, the richness of its wildlife is staggering. In the early mornings thousands of doves flock in to drink at the water-holes, the piping whistle of their wings drowning out all other sounds. They are followed by hundreds of sandgrouse – the Namaqua sandgrouse with its melodious 'kelkiewyn' call and Burchell's sandgrouse with its distinctive 'choclit, choclit' cry. At any moment the drinking flocks can erupt into a frenzy of evasion as a lanner falcon swoops in on them, usually to miss, although now and then a puff of feathers left floating silently in the still desert air marks a kill.

At certain times of the year large herds of gemsbok may be seen in the dry river-beds or posing on the red sand dunes providing one of the Kalahari's most evocative images. During summer evenings the barking geckos set up a chorus around the Nossob and Mata Mata camps against a background of churring nightjars while, on occasion, you may hear the whoop of a spotted hyena – the most exhilarating sound of the African night. No-one will be able to forget being woken in the early mornings by the grunting roar of lion carrying across the silent desert.

LEFT: *When the wind blows in the Kalahari, large dust devils wind across the river beds and pans, throwing palls of dust high into the air.*

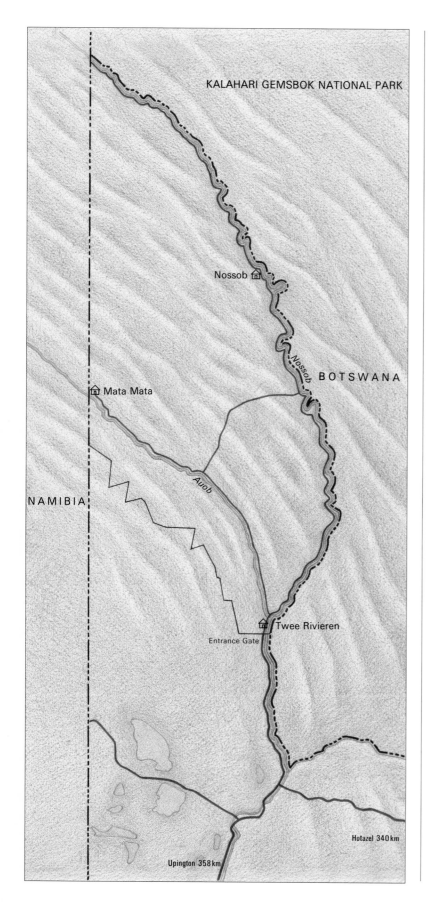

A watchful suricate gathers her young protectively around her at the entrance to their burrow. They live in the ground for protection and for shelter from the extremes of temperature.

The Kalahari is renowned for the flocks of migratory eagles and falcons that visit in summer to feed on flying termites and ants. Then there are the irregular but spectacular mass migrations of game such as eland, blue wildebeest, red hartebeest and springbok, of which the most recent took place in 1987.

The crucial factor as far as the wildlife of the area is concerned is the sheer size of this conserved area. Although the Kalahari Gemsbok National Park covers a substantial 9 600 square kilometres inside South Africa, to this is added a further 26 600 square kilometres of Botswanan territory known as the Gemsbok National Park. There is no fence between the two parks and there is a high degree of cooperation between the South African National Parks Board and the Botswanan authorities. The combined total of 36 200 square kilometres (3,62 million hectares) makes this conservation area considerably larger than the Kruger National Park which is just under 20 000 square kilometres (about 2 million hectares) in extent. (Belgium, by comparison, has a land surface area of 30 513 square kilometres.)

All this space is needed because the game animals of the Kalahari are adapted to moving in response to localized rain patterns in order to find the best grazing conditions in this region where the annual rainfall averages as little as 200 to 250 millimetres. What is more, the precipitation is erratic with periods of plentiful rain alternating with devastating droughts.

The key to the permanent survival of this wilderness and its wildlife is to ensure no fence is ever erected in the bed of the Nossob River which forms the international boundary between South Africa and Botswana.

There is no permanent surface water in the Kalahari but under 'normal' conditions the various antelope species can survive without surface water, obtaining all their moisture requirements from their food plants. In drought years the plant crops may fail, there is no moisture in the grazing, and the great herds migrate or 'trek'. There is evidence to suggest that in the past they may have ranged as far south as the Orange River in their search for green pastures but that is no longer possible because of the fences now crisscrossing the farming areas of the northern Cape and southern Botswana. The numerous boreholes sunk by the Parks Board in the dry beds of the Nossob and Auob rivers provide the precious water sought by the migrating herds and, in the winter of 1979, they helped to sustain the estimated 85 000 to 170 000 blue wildebeest, 5 000 eland and 6 000 red hartebeest which trekked through the park.

Another trek took place in 1985 when, yet again, thousands of eland, wildebeest and red hartebeest from Botswana entered the northern section of the Kalahari Gemsbok Park over a front of hundreds of kilometres. The fence on the western boundary with Namibia prevented the game moving westwards out of the park and the mass of animals moved southwards, exhausting the water at the various waterholes. According to park warden, Elias le Riche, the windmills could not cope with the demand and diesel-powered pumps had to be installed at 30 boreholes to provide the required volumes of water. In spite of this hundreds of animals died of starvation.

Amongst the trekking eland were large numbers of calves abandoned because they could not keep up with their mothers and which, by the time the herds reached Twee Rivieren at the south end of the park, seemed doomed. Many were saved by being captured and supplied with food and water, and were later translocated to other reserves in South Africa. The migrating herds eventually moved in a circular pattern eastwards and northwards back into Botswana.

Observations over the years have led researchers to conclude that, in overall terms, the tree savanna areas with a high proportion of annual grasses in the north-east of Botswana's Gemsbok Park attract many of the grazing antelope. However, when drought strikes this region, the perennial grasslands in the south-western parts of the combined park, that is, South Africa's Kalahari Gemsbok National Park, are crucial for their survival.

Such spectacular mass movements of game are rare and lucky the tourist who may happen to witness them. By and large, visitors to the park will have to settle for less dramatic sightings. This is not the Kruger National Park – there is less 'big game'

both in numbers and diversity of species – and because the area is so arid the animals are often more widely spaced. Yet there is so much to see that is different from what one is used to in the more conventional game reserves situated in lusher lowveld and bushveld areas.

The base upon which the Kalahari ecosystem is built, and the lowest level of the food-pyramid leading up to the large carnivores at the top like the lion and the eagle, is the region's drought-adapted plantlife. This consists of perennial plants which provide food and shade for animals throughout the year and the annuals which complete their life-cycle from germination to death inside a year. Many of the annuals are programmed to respond to the region's sparse rainfall so that they germinate, flower, produce seed and die within a month of a rainstorm, after which the seeds lie dormant in the soil until the next rains.

Among the region's most interesting and important plants are two creeping cucurbit vines which bear succulent fruit – the tsamma melon which is an annual and the gemsbok cucumber which is a perennial. The tsamma occurs in patches and its fruits look like a collection of striped gem squashes. The cucumber fruit is about a quarter the size of a tsamma, is greeny-yellow when ripe and is covered in spines. The cucumber also has a thick, succulent root which antelope dig out for its sap.

The seeds of these plants make nutritious food but their main value is as a source of moisture because about 95 per cent of the

A pale chanting goshawk dangles its prey. It can prove to be a formidable predator, taking mainly snakes, lizards, insects, and small birds and mammals.

A Burchell's sandgrouse soaks its belly feathers to take water back for its nestlings to drink.
RIGHT: *Despite their formidable array of quills, porcupines are frequently taken by lions who often suffer painful injuries in the process.*

fleshy fruit consists of water. They are eaten by an enormous array of wild creatures, from insects like the armoured ground cricket, to birds, rodents, most antelope and even predators such as the brown hyena and honey badger. The cucumber produces a crop of fruit every year but the tsammas can fail completely during the drought years.

The Kalahari's perennial grasses include species such as the small bushman grass which dominates the riverine plains, particularly along the banks of the Nossob. This grass has a white, feathery flower head and it is a striking scenic feature where it occurs in extensive stands. However, the grass which can really transform the Kalahari is an annual known as 'suurgras' or Kalahari grass. After good rains it grows profusely to form tall, thick stands about a metre high. The contrast between the bare winter Kalahari landscape and one with swards of 'suurgras' extending to the horizon is quite astonishing. Herbivores are able to eat 'suurgras' when it is young but as it matures it produces a sticky, acid secretion which causes open sores on their tongues. This defensive mechanism forces grazers to leave the grass alone until it dries out at the end of summer, when it can safely be eaten again.

The dominant tree along the river-beds of the Kalahari is the camel-thorn which occurs mainly along the dry river-beds where the park's roads run. Some of these trees grow to a height of 15 metres, drawing water from their extensive root systems which penetrate deep below the river-beds. Many appear to

have been 'thatched' with dry grass. These haystack-like masses are the colonial nest structures of the sparrow-sized sociable weaver and may provide separate nesting chambers for up to 50 pairs of birds. The diminutive pygmy falcon may often be found nesting in appropriated chambers of the colony. The camel-thorn fulfils a multitude of important functions in the Kalahari ecosystem. Its leaves and pods are eaten by a wide range of animals, it provides shade for a variety of larger mammals, while large raptors build their nests in the crowns, and spotted and giant eagle owls roost in them by day.

Unfortunately a large number of camel-thorn trees were destroyed in the Nossob valley in 1976 by raging veld fires caused when lightning flashes ignited the dense grass cover that

had developed after several years of good rains. Nearly half of all the larger trees were destroyed and a management priority in the park since then has been to keep uncontrolled fires that originate in Botswana's section of the combined park out of the Nossob. Natural fires, however, do play an important part in maintaining the ecosystem.

Perhaps the commonest tree in the dunes is the shepherd's tree or 'witgat' which is also extremely important as a food source for a wide variety of animals. In addition to providing food both the shepherd's tree and the camel-thorn also provide shade – but at a price, for living in the sand below them are colonies of hundreds of sand tampans. These small ticks are alerted by shifts in the sand caused by large animals moving around near them. They make their way to the surface of the ground where they locate their prey by using carbon dioxide receptors that 'zero in' on the exhaled breath of the animal.

The tampan sucks blood from the animal, after first anaesthetizing the site of the bite with a neurotoxin secreted from its mouthparts. Antelope appear largely unaffected by sand tampans but domestic stock and humans often develop painful, suppurating sores from the bites which can have serious consequences. In extreme cases sand tampans may be numerous enough to cause stock to die from loss of blood.

Most visitors are interested solely in large game species and, in particular, finding cheetah and the magnificent Kalahari lion. However, the Kalahari teems with smaller animals superbly

The constant procession of doves arriving to drink at the waterholes are easy prey for the African wildcat.

adapted to living here and coping with ground surface temperatures that can reach 70 degrees Celsius on a hot summer's day and minus 10 degrees Celsius on a freezing winter's night. These include such readily seen species as the ground squirrel, suricate, yellow mongoose and bat-eared fox.

The mechanisms developed by these animals to cope with life in this harsh environment are both numerous and fascinating. The ground squirrel, for example, which has taken up residence inside all three rest-camps, protects itself from the intense heat of the sun by using its bushy tail as an umbrella to provide much-needed shade while it feeds.

The suricate, a particularly engaging species of mongoose, forages through the veld in highly organized social groups looking for its mostly insect prey. They also eat lizards and scorpions and, like other mongooses, are capable of killing venomous snakes. Each band posts a full-time 'sentry' to make sure that they do not fall prey themselves to any of the park's numerous eagles. The bat-eared fox uses its huge ears to pinpoint the almost imperceptible sounds made by its principal prey of termites and beetles and their larvae, while the Cape fox uses its luxuriant bushy tail as both a counterbalance for its body while swerving at high speed and as a distraction to fool pur-

suers as to which direction it intends turning. All of these animals live in burrows in the ground for protection and for shelter from the extremes of temperature.

The antelope from which the park derives its name, the gemsbok, is admirably adapted to this environment in a number of ways, to the point where it appears to drink water for the minerals it contains rather than to obtain moisture. That became apparent from a project in which several series of adjacent windmills in the Auob river-bed were closed down for a year at a time between 1978 and 1982. According to Dr Gus Mills, who was the research officer in the park for 12 years, the results showed gemsbok were definitely attracted to the closed windmills because of the higher mineral content in the soil left behind from the evaporated borehole waters.

The study also showed that although the presence or absence of water had no effect on the movements of springbok, it had a major influence on the blue wildebeest: of all the Kalahari antelope, this is the most dependent on regular access to water.

RIGHT: *The heavily maned Kalahari lion is often believed to be larger than lions found elsewhere. The apparent size difference, however, is an optical illusion created by the wide-open Kalahari landscape.*

Brown hyena are mainly nocturnal and are most often seen in the early morning and late evening.
LEFT: *A pride of five cheetah finish off a springbok. The Auob River is one of the best places in the park to find cheetah.*

Research on captive gemsbok has shown that they have the amazing ability, under conditions of extreme heat when the temperature rises to 45 degrees Celsius, to allow their own body temperature to rise to more than 45 degrees for up to eight hours if necessary. They do not sweat as by doing so they would lose valuable moisture. Although most animals would succumb to brain damage if their body temperatures reached levels as high as 45 degrees, the gemsbok survives because it has a special system of blood vessels below the brain called the carotid rete.

The carotid artery, carrying blood from the heart to the brain, breaks up into a network of vessels before it reaches the gemsbok's brain. This system of arteries is surrounded by veins returning cool blood to the heart from the gemsbok's nasal sinuses. That blood has been cooled by the animal's panting and, operating like a heat exchanger, the system cools the blood in the carotid artery by as much as three degrees, thus protecting the brain from excessive heat.

The gemsbok, incidentally, is able to kill a lion with its rapier-like horns and the combination of this armament and the animal's speed earns it considerable respect from the Parks Board game-capture teams. Comments one member: 'You could wind up skewered like a sosatie if you don't watch out!'

The graceful springbok, according to Gus Mills, possibly has a higher rate of reproduction than any other antelope. Every adult female produces a lamb each year no matter how poor the conditions and may fall pregnant again within days of lambing to produce two lambs in a 13-month period. Under particularly good conditions seven-month-old immature females may

Many of the annuals found in the Kalahari are programmed to respond to the region's sparse rainfall so that they germinate, flower, and produce seed within a month of a rainstorm.

Tsamma melons provide a much needed source of water for an enormous array of wild creatures through much of the long, dry season.

become pregnant and will give birth six months later at just over one year when they are not yet themselves fully grown.

This fecundity may lie at the root of the periodic treks of hundreds of thousands of springbok recorded by naturalists last century when huge springbok populations built up after years of good rains and abundant food. The last two treks in the southern Kalahari took place during the 1940s and 1950s. Springbok ('jump-buck') are named for their habit of 'pronking'. This extraordinary display occurs when an adult is alarmed or when lambs are at play; it makes a series of stiff-legged jumps, landing on all four hooves at once and bouncing into the air again. Its head is held low while jumping and a crest of long white hair along its lower spine is fluffed out into a spectacular fan. A pronking springbok is displaying its fitness to predators in an attempt to discourage them from attack.

The heavily maned Kalahari lion is often believed by visitors to be larger than lions found elsewhere but, according to Mills, this is not the case. It just looks that way because of the openness of the landscape. Warden Elias le Riche estimates total lion numbers at about 400 with about 150 of these found in the South

African section of the combined park. A particular concern of his is the large numbers – up to 40 – that annually leave the unfenced south-eastern border of the park inside Botswana and are shot by farmers in the area because they turn to killing cattle. The Parks Board plans to extend the border fence by about 70 kilometres along this stretch to fence off the commercial ranches in Botswana bordering the park.

Research by Professor Fritz Eloff has shown that the Kalahari lion leads a tough life. Although the gemsbok is its most important prey, more than 50 per cent of the total kill is made up of much smaller animals, particularly porcupines. Lions have to survive on many little kills when the larger antelope disappear on treks or are otherwise hard to find. Most lion prides do not follow the migrations of the antelope but remain in their established territories which may be up to one thousand square kilometres in extent. According to Eloff the main factor controlling the lion population is the death rate amongst cubs. Statistics for the rest of Africa show two-thirds of all cubs die before they reach one year of age but the death rate for cubs in the Kalahari is even higher.

The cheetah is the other predator which most visitors would like to see and, again according to Mills, the stretch of land along the Auob River is one of the best places in the world to watch them hunting. The narrow river-bed concentrates the cheetah's main prey, springbok, and as many as four female cheetah with their cubs as well as several males live along the 130-kilometre stretch of 'river' between Twee Rivieren and Mata Mata. Finding cheetah requires slow driving and keen eyes, while witnessing a kill is one part luck and nine parts patience, staying with the cheetah once you have located them.

Much of the activity in the Kalahari takes place at night and it is only thanks to the efforts of researchers like Mills and Eloff who have followed predators around at night or followed their tracks in the sand during the day that we have a picture of the habits of animals such as the brown and spotted hyena. These are rarely seen by visitors and then usually very early in the morning before they retreat to their dens for the day.

If ever an animal has suffered from a bad press it has been the spotted hyena. Once believed to be a cowardly scavenger it is, in reality, the dominant predator in the Kalahari after the lion. As Mills comments, 'How can an animal that has one of the most highly developed social systems amongst carnivores be called stupid; one that can run at up to 50 kilometres per hour for up to three hours be called ungainly; or one which is bold enough to chase lions into trees be called a coward?'

Brown hyena are more common than spotted hyena but are just as rarely seen. Mills has studied them intensively and radio-tracking has shown them to be 'loners' travelling on average 32 kilometres a night as they prowl for food.

The complex interactions between predators and prey in the Kalahari have been closely studied, revealing many interesting relationships and habits. Spotted hyena and lion prey to a large extent on the same species but research has shown that 83 per cent of the antelope killed by lions are adults or subadults compared with only 42 per cent for spotted hyena which tend to take mainly juveniles. In this way competition between the two predators is reduced.

The young of the various antelope species are obviously vulnerable to predation but springbok, for example, cope with this by synchronizing the births of their lambs to a marked degree. The springbok's many predators are temporarily 'swamped' with easily caught prey for a few weeks, and are satiated after catching only a small percentage of the total lamb crop. After those first few hazardous weeks the surviving young are fast enough to look after themselves.

The park was proclaimed in 1931 thanks to the efforts of Piet de Villiers, Inspector of Lands at Upington, and Willie Rossouw, a local farmer. The latter was the son-in-law of Christoffel le Riche, a pioneer who trekked into this area around 1880. His descendants have been intimately involved in running the Kalahari Gemsbok National Park since its inception.

De Villiers and Rossouw brought the large-scale destruction of game in the area by hunters to the attention of Minister of Lands Piet Grobler, who had played a key role in the establishment of the National Parks Board and the Kruger National Park, in 1926. Grobler became equally concerned about the fate of the wildlife of the Kalahari and it was through his vigorous campaigning and political influence that the Kalahari Gemsbok National Park was finally proclaimed. Botswana's Gemsbok National Park was proclaimed in two stages in 1938 and 1972.

The first ranger to be appointed was Johannes le Riche, son of Christoffel, and since then the post of warden has remained in the Le Riche family, passing to brother Joep, and then in turn to Joep's sons Stoffel and later Elias, the present warden. When he was asked why the region is so attractive to himself and to his family, Elias le Riche commented simply: 'I like the climate and I like dry country.'

Poaching remains one of his major problems, with large-scale incursions taking place regularly in the Botswanan section of the combined park. Le Riche has to cover this vast territory with just 18 rangers and he feels that he could do with double this staff complement. Poachers caught in Botswana are handed over to the Botswanan police to be dealt with.

The best time to visit the Kalahari Gemsbok Park is during March to April. The area usually receives most of its rainfall during January and February, frequently in the form of violent

Lappetfaced vultures dominate at kills and this one is clearing a space for itself amongst the more numerous, but smaller, whitebacked vultures.

Of all the Kalahari's large antelope, research has shown the blue wildebeest to be the most dependent on regular supplies of water.

thunderstorms which can be exciting, if a little frightening to experience, particularly if one is camping. The gusts of wind blowing ahead of the storms may break branches off trees and, in extreme cases, are powerful enough to uproot large camel-thorn trees and bowl over windmills.

The build-up to the rains can be depressingly oppressive. Weeks pass with the daily maximum temperature constantly above 35 degrees with no sign of relief. When the winds blow it feels like a hot gust from an oven and large twisters or dust devils wind across the river-beds and pans throwing palls of dust high into the air.

Then, the wind turns to the north, the nights seem a little more humid and the first powder-puff clouds appear in the sky. It can take another tense week before they build up into the great blue-black thunderclouds which produce the life-giving rain.

After the rains much of the game moves into the river-beds where they are easily seen from the tourist roads. They remain there until the first frosts in early May after which most of the animals begin to move out into the dunes. Also in late summer the park attracts hordes of migrant raptors to augment its already impressive resident raptor population. Lanner falcon, for example, are resident in the park and occur normally in pairs but it is not unusual to see flocks of up to 15 of these birds feeding on flying-ant swarms during summer when the abundant food supplies attract immigrants.

The park is particularly rich in owls and at Nossob Rest-camp visitors can regularly find scops, pearlspotted and whitefaced owls roosting in the trees during the day. The impressive giant eagle owl can usually be found by searching large camel-thorn trees around the water-holes near Twee Rivieren and Mata Mata rest-camps.

Whichever route is chosen to reach the park, the last 340 kilometres or so from either Hotazel or Upington is over gravel roads whose condition depends largely on when they were last graded. Any car in good condition can cope with the journey but there will be a certain degree of wear and tear. In particular, travellers will find out just how good their car's dust-proofing is and may expect some attrition on tyres. Graders work on these roads constantly and their operations can cause problems for low-slung vehicles with little ground clearance, particularly if they are forced to cross over the 'middel-mannetjie' or hump built up in the road by the grader's blade. Care must be taken because the sand hump often contains rocks. The road north from Upington is currently being tarred.

The roads inside the park follow the two river-beds of the Nossob and Auob except for the two connecting roads which are built between them across the dunes. The second dune road was under construction at the time of writing. These roads are usually kept in very good condition but drivers should be careful not to become stuck during the rains when washaways occur in sections.

The two rivers are dry except for short periods when they will flow temporarily after particularly good rains. On average the Auob flows about twice a decade and the Nossob about twice a century. Temporary pools of water form in parts of the river-beds after normal rains.

Should vehicles become stranded or break down, the golden rule is to stay with the car. Travellers are logged in and out of the rest-camps and, if another visitor does not come to the rescue, park staff will set off to find the missing vehicle if it has not reached its stated destination by gate-closing time.

Sections of the roads are sandy and it is possible to get stuck if a stop is made in the wrong place. It is usually possible to extricate one's car by deflating the tyres by up to 50 per cent. This increases the vehicle's traction and, once out of the soft sand, no damage will be done driving on the deflated tyres as long as the speed is kept down.

Accommodation facilities in the park have recently been expanded and upgraded and, according to Le Riche, future developments will depend on whether the access road from Upington is eventually tarred as this will make it easier for larger numbers of tourists to visit the Kalahari. That being the case, Le Riche says the Parks Board plans to upgrade the stand-ard of roads within the park and to build a number of satellite camps around Twee Rivieren. These would utilize the existing administrative infrastructure and facilities at the main camp such as the restaurant and shop. A number of bush-camps would then be built in the dunes, which visitors would be able to hire on an exclusive basis. In addition, the accommodation at Nossob and Mata Mata camps would be upgraded although no extra huts can be built there because shortage of water restricts the number of people each camp can support.

Other plans include developing special routes and camps for four-wheel-drive vehicles in the Botswanan section of the combined park, subject to negotiation with the Botswanan authorities, and also to set up three-day safaris through the desert using camels as transport.

Le Riche says the Parks Board has no intention of tarring the roads inside the park itself, commenting: 'The Kalahari would not be the Kalahari without dust on the roads.'

The formidable horns of the gemsbok are a distinguishing feature as they stand silhouetted against the late afternoon sun.

AUGRABIES FALLS NATIONAL PARK

Originally set up in 1966 to conserve the wildlife and spectacular landscapes surrounding the Augrabies Falls where the Orange River plunges over a 56-metre granite cliff into an 18-kilometre long canyon, the Augrabies Falls National Park now has another, perhaps equally important, role – as a conservation area for one of the most endangered of the four subspecies of the black rhinoceros, the Cape rhinoceros.

Augrabies is in many ways ideal for black rhinoceros and the National Parks Board has been working over the past few years to meet the requirements for a large population to be built up here. If all goes according to plan then in thirty or so years time Augrabies could be associated with black rhinoceros the way that the Mountain Zebra and Bontebok national parks are associated with those two mammals they saved from the brink of extinction.

Black rhinoceros occurred in the Augrabies area before they were shot out in the first half of the 19th century, with the last known animal in the Cape Province being killed in 1853 between Uitenhage and Addo. In addition to providing good habitat Augrabies meets one other essential requirement for conserving black rhinoceros – this remote, harsh and underpopulated region is a relatively safe haven for them because of the difficulties it poses to potential poachers.

The Parks Board has also enlisted the help of the South African Defence Force (S.A.D.F.) through the agreement which made the adjoining 70 000-hectare Riemvasmaak military training area a part of Augrabies as a 'contractual national park'. In terms of this agreement, the wildlife resources of the area are protected and managed as if the area were a full national park, but the S.A.D.F. retains ownership and will continue to use sections of it as an artillery and mortar firing range and as a bombing range by the Air Force. Around 20 000 hectares of infantry-training ground have been excluded from the contract and will be fenced off from the park.

LEFT: *On its journey to the Atlantic Ocean, the Orange River plunges over the Augrabies Falls into the deep gorge below them.*

The technicoloured Cape flat lizards are abundant in the area and are frequently seen on the walk along the edge of the Falls.

The catastrophic decline in Africa's black rhino population over the last 15 years has been well documented and the mammal has become a 'flagship' species for the conservation movement. Like the world's four other rhinoceros species, the black rhinoceros – *Diceros bicornis* – has been slaughtered by poachers for its horns. These are prized for various traditional medicinal properties in some Far Eastern countries such as Taiwan and China while in the Yemen Arab Republic rhinoceros horn is the material traditionally used for the hilts of ceremonial daggers.

Conservationists estimate that Africa's total population of black and white rhinoceros has been slashed from around 100 000 in 1960 to only 8 240 in 1992, simply from poaching pressure. According to statistics and estimates compiled by the African Rhino Specialist Group of the World Conservation Union (I.U.C.N.) black rhinoceros numbers in Africa have tumbled from about 14 800 to just 2 490 in 1992.

The largest remaining black rhinoceros population left in the wild is now found in Zimbabwe, but here too, poaching pressure has intensified over the last few years. Pitched battles are being fought between poachers and game rangers who are operating under orders to shoot to kill poachers on sight.

That any black rhinoceroses remain in South Africa at all is thanks to the Natal Parks Board because by the late 1940s the only black rhinoceroses left in the country were about 150 or so in three of Natal's game reserves. At that time, however, efforts were being concentrated on saving the southern white rhinoceros which was then on the brink of extinction. No-one at the time was particularly concerned about the Natal black rhinoceros because as there were so many of them left in the rest of Africa the population did not seem to be threatened. Nevertheless, Natal looked after its black rhinoceroses well and was able to distribute them to other areas, in particular to the Kruger National Park.

According to Daryl and Sharna Balfour in their book *Rhino – The Story of the Rhinoceros and a Plea for its Conservation*, a national

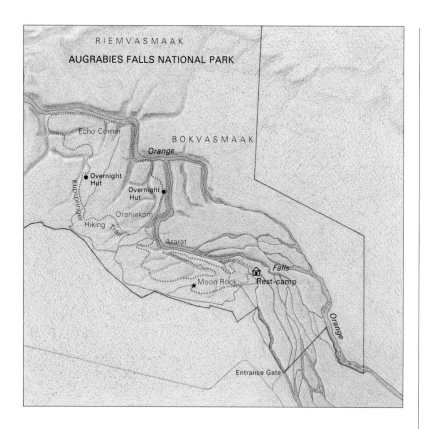

again exchanging some of its surplus game of a variety of species for the rhinoceros, a cow named Shibula. She had originally been sent from Namibia as a mate for a bull in the Lisbon Zoo, but when the latter died and no replacement could be found, Shibula was shipped back to Africa in the greater interests of the conservation of the subspecies *bicornis*. Within a few months of her arrival at Augrabies it was apparent that she had successfully adapted to the diet provided by the park's vegetation.

From these small beginnings it is certain that the rhinoceros population will grow steadily. The enlarged park, taking in the Riemvasmaak area, is estimated to be capable of supporting up to 100 black rhinoceros compared with only a dozen or so in the original north bank section. The Riemvasmaak area is not yet open to the public, however.

The Augrabies Falls themselves are overshadowed in a southern African context by the 100-metre-high magnificence of the Victoria Falls in nearby Zimbabwe, but are still a sight worth seeing – in part because of the sharp contrast afforded by the great cascades of water pouring through the parched and barren landscape of one of the most arid regions of South Africa. The Augrabies Falls gets its name from the Nama word 'aukoerebis' meaning 'place of great noise'. The average annual rainfall at Augrabies is a mere 107 millimetres and summer temperatures soar into the mid-40s Celsius.

The Orange – South Africa's largest river – rises in the Drakensberg range some 15 kilometres south of Mont-aux-Sources and flows 2 340 kilometres westwards until it reaches the Atlantic Ocean at Alexander Bay.

Modern man, however, has done much to tame and control the natural flooding cycles of the Orange and this has had a direct effect on the amount of water passing over the Augrabies Falls. Very rarely nowadays is there a peak flow of the sort that makes the Victoria Falls on the Zambezi such a marvellous spectacle. The Orange has been tamed by engineers and water released from the two massive reservoirs – the Hendrik Verwoerd and P.K. le Roux dams – is maintained at a more or less constant flow where possible. Only when they are full can the flood waters of the Orange River surge westwards towards Augrabies and the sea.

Ironically, when the Orange really does experience peak flooding, as it last did in 1988, visitors cannot see the effects at Augrabies because the national park has to be closed. The river spreads itself as much as six kilometres across, cutting off the

strategy for the black rhinoceros has now been drawn up in which Augrabies is destined to play an important role. 'The Conservation Plan for the Black Rhino, *Diceros bicornis*, in South Africa, the Homelands States and Namibia' was drawn up by Dr Martin Brooks of the Natal Parks Board for the Rhino Management Group which is made up of one member from each of the conservation bodies in these countries which have rhinoceros under their control.

There are two separate subspecies of black rhinoceros recognized for South Africa and Namibia. *Diceros bicornis minor* occurs in Natal and the Kruger National Park, and the subspecies which previously occurred in the northern Cape to western Namibia is *Diceros bicornis bicornis*. In 1961 and 1962 seven rhino of the East African subspecies *michaeli* were purchased from the Kenyan authorities and placed in Addo Elephant National Park. Their descendants now number more than 20. Ironically enough, when Kenya sold the rhinoceros to South Africa its own population was satisfactorily conserved: now that poaching has decimated East Africa's rhinoceros, the *michaeli* at Addo have become crucially important to the future of this subspecies.

Augrabies rhinoceros came from Namibia's Etosha National Park in exchange for buffalo from Addo Elephant National Park (valued by conservationists because they are guaranteed free of foot-and-mouth disease). A recent additional black rhinoceros came from Lisbon Zoo in Portugal and was acquired through international conservation diplomacy with the Parks Board

RIGHT: *The Moon Rock is a distinctive landmark which acquired its shape through erosion. While the inner temperature of the rock remains stable, the outer layer expands and contracts according to temperature fluctuations which eventually results in it peeling away to form a smooth dome.*

The Augrabies Falls are a magnificent sight when the Orange River is in full flood.

Dainty, ballerina-like klipspringer blend in with the granite rocks that form their preferred habitat.

rest-camp and pouring into the ravine through a series of lesser waterfalls in addition to the main cataract. In 1988 three kilometres of the access road to the national park were washed away by flood waters as was the suspension foot bridge linking the camp on the south bank to Riemvasmaak on the north bank. As of 1992 the foot bridge had not been replaced, and park staff still have to undertake the 80-kilometre round trip by road to reach Riemvasmaak.

Above the Falls, the Orange forms what geographers call a braided stream, with its water channelled between numerous islands built up from the deposition of alluvium by the slowly moving river. It is fortunate that part of this landscape feature falls within the park because a distinctive riverine vegetation has developed over this alluvial material – totally different in composition and appearance from the desert-adapted 'Namaqualand Broken Veld' veld-type so characteristic of the rest of the park. As is readily apparent to visitors travelling from Upington to Augrabies, virtually all of this natural riverine vegetation along the banks of the Orange has been cleared to make way for irrigated crops, particularly grapes. And thus the national park can be credited with saving a vitally important fragment of one of South Africa's most seriously threatened plant communities.

The Falls were formed between 500 and 600 million years ago when the tectonic forces shaping the subcontinent pushed up a huge section of the landmass to form an interior plateau. Instead of flowing gradually westwards towards the Atlantic as before, the Orange now poured over the edge of the plateau, cutting into the bedrock until it found a weak spot – or nick point – in the rock. The river now gouged harshly into this point, eating it away and lowering it, and so bringing more and more of the full power of the flow to bear until it eventually carved out the ravine visitors can see today. The waterfall is still – if imperceptibly – eroding its way eastwards.

Human control over the Orange will increase early next century with the completion of the Lesotho Highlands Water Project. This scheme will siphon water from the upper reaches of the Orange River in Lesotho and divert it to the Vaal River. By effectively doubling the annual flow into the Vaal basin, it will ensure that another of South Africa's huge reservoirs – the Vaal Dam – is able to continue supplying the ever-growing water needs of the country's industrial heartland situated in the Pretoria-Witwatersrand-Vereeniging triangle.

Visitors to Augrabies in the summer months are likely to bring away two particular memories in addition to the scenic and other splendours of the park – the extreme heat and the ubiquitous black fly. The latter can be a great nuisance, swarming around faces and trying to crawl into noses and ears.

These tiny midge-like flies, of the genus *Simulium*, are found worldwide in association with rivers and, while the males eat only nectar, the females of the species are bloodsuckers on birds or mammals. They are somewhat less considerate of their unwilling hosts than that other notorious vampire of the insect world, the mosquito: one American outdoorsman has claimed that whereas the mosquito uses the equivalent of a hypodermic syringe, the black fly's biting equipment resembles more a chain saw. Fortunately for visitors to Augrabies, the local species does not bite man. It can, however, be an unmitigated nuisance and an insect repellent spray or cream may be necessary on occasion.

The flies have an interesting lifecycle. The females lay their eggs on underwater stones and, on hatching, each larva attaches itself to a stone by a sucker at its tail end to avoid being washed away by the fast current. In due course the pupa is also formed underwater, also attached to a stone. When the adult fly is almost ready to emerge it stores air inside the pupa so that when the pupal case ruptures the fly is shot to the surface in a bubble of air and immediately flies off.

By building dams on the Orange River man has unwittingly increased the black fly problem by creating a more regular flow of water and thus providing a more stable environment for black

The distinctive kokerboom stands alone in the broken, rocky country that makes up much of the Augrabies landscape.

Droplets of dew glisten on this species of Boraginaceae *which is often likened to a weed.*

Rosyfaced lovebirds are a common species of the dry, western areas, often gathering in noisy flocks to drink at waterholes.

fly larvae. One method of control is to reduce the flow of the river temporarily. This lowers the water level and exposes the rocks to the sun, killing the pupae outright. The larvae are forced to release their hold on the rocks and drift off in the stream where large numbers fall prey to fish.

Perhaps the most common birds in the park are the flocks of palewinged starlings which are often seen around the rest-camp and Falls area. They feed on a wide range of foods from insects, nectar and fruit to table scraps around picnic sites, and can become quite tame. That most agile of antelope, the klipspringer is, also easily seen in the park on the hiking trails or in the vicinity of landmarks like the Moon Rock and Ararat. Like the palewinged starling it has also become rather tame. The fortunate visitor may witness an interesting interaction between the two when the starlings clamber over the unresisting klipspringers in search of ticks or other ectoparasites. The starlings here have taken over the role of the two oxpecker species which do not occur in the arid western regions of southern Africa.

Klipspringers are readily seen because of the male's habit of posing, sometimes ballerina-like with all four hooves close

together, on exposed rocky outcrops. This is in fact a territorial display to other males and it is reinforced by scent-marking using secretions produced by a gland on either side of the klipspringer's face below its eyes. The klipspringer pushes the gland opening against the end of a bare twig to smear the black tarry substance onto its tip.

The birdlife at Augrabies is of course limited by the aridity of the region, but is nonetheless interesting for the discerning observer. Around the gorge during the summer months, for example, thousands of alpine and black swifts arrive to breed on the sheer cliffs while the rosyfaced lovebird is an Augrabies 'special'. There is usually a breeding pair of peregrine falcons – the fastest birds on earth as they chase the swifts against the backdrop of the granite gorge and the turbulent muddy waters below. Another unusual species sometimes seen in the gorge or elsewhere along the cliffs that border the Orange River is the black stork. It nests on the cliffs in the winter months.

Away from the river the landscape of the park is dominated by rugged, rocky hills although there are occasional huge, low, dome-shaped rocks such as the Moon Rock. These have been formed by exfoliation, where the outer 'skin' of the rock expands and contracts with the extreme daily variations in temperature until it eventually cracks and peels off, layer by layer like an onion. Much of the park's bedrock is made up of gneiss which has a characteristic orange or pink-orange colour in the early morning and late evening light. This pattern is dramatically broken by the low granulite outcrops which stand out stark and black against the orange/brown background.

The vegetation of the park is adapted to coping with the prevailing drought conditions, the extremes of temperature and the poor soil. Shrubs such as the *Namaqua cereria* are able to grow in soil blown into small fissures or cracks in solid rock. Some plants cope with drought through having evolved mechanisms which conserve their water, but others avoid drought altogether by germinating, flowering and producing seed all within a month of a shower of rain. One example of this at Augrabies is a tiny aquatic plant, *Limosella capensis*. It grows in the shallow rock pools which form on top of the granite domes after rain and which hold water for a few brief weeks.

The most distinctive plant in the park is the giant aloe known as the quiver tree or 'kokerboom' (*Aloe dichotoma*). It was so named because the bushmen used the branches to make quivers for their arrows. This aloe is found throughout the northern Cape and southern Namibia and its distinctive silhouette typifies the rugged, barren landscapes of these regions. The quiver tree flowers during the winter months when the pale-winged starlings around the Augrabies rest-camp will often be seen with yellow foreheads – dusted with the pollen from the quiver tree flowers on which they have been feeding.

While the Falls themselves will probably remain the main attraction for the average visitor to Augrabies Falls National Park, there can be little doubt that its role as a black rhinoceros sanctuary will increasingly bring this arid but beautiful park to the attention of conservationists worldwide.

When agitated, springbok are renowned for performing spectacular vertical jumps known as 'pronking'.

KAROO NATIONAL PARK

Although the Great Karoo of South Africa covers nearly a quarter of the country's land surface area, it used to be one of our least-understood and least-studied regions. To the man-in-the-street it is perhaps best known as the longest and dreariest stretch of the 1 400-kilometre-long road journey between Johannesburg and Cape Town. There seems to be no end to it and the tendency of many motorists is to cruise at speeds beyond the legal limit as they cover the hundreds of kilometres of road running arrow-straight to the horizon. For them the grey monotony of the Karoo is broken only by the occasional farmhouse and odd windmill and the only

Sparse thorn bushes are the favoured habitat of the Karoo dwarf chameleon. When threatened, it loses its bright colour and will attempt to conceal itself amongst foliage.

signs of 'wildlife' they see are the road kills – squashed hares, eagle owls, polecats and bat-eared foxes. These unfortunate creatures are run over at night by drivers not prepared, or too tired, to slow down and give the light-dazzled animals a chance to get out of the way.

And yet, if these travellers would just slow down and pause to take a closer look they would find that, like so many other barren semi-desert regions, the Karoo plays host to a remarkable array of wildlife riches. It also has a relaxed and easy-going atmosphere that could lead the traveller to question why he is so intent on hurtling as fast as possible from one concrete jungle on the Witwatersrand to another one, albeit more scenic, on the western Cape coast.

For many years, however, opportunities for the tourist to learn more about the Karoo have not been available. Only in the last 15 years or so have steps been taken to conserve sections of Karoo vegetation with the establishment of the Tankwa Karoo and Karoo national parks as well as the Karoo Nature Reserve at the Valley of Desolation near Graaff-Reinet.

The proclamation of the Karoo National Park near Beaufort West in 1979, followed by the construction of a luxury rest-camp

LEFT: *The eternal Karoo landscape – wide open plains stretch to an endless horizon punctuated by flat-topped koppies.*

in 1989, now offers the visitor a chance to discover the Karoo and the accommodation statistics for the park show that more and more tourists are doing just that. Given the park's close proximity to the N1 highway, it is perhaps not surprising that at present about 80 per cent of the accommodation is occupied by travellers who are merely using the park as a convenient overnight stop on their way to or from Cape Town and Johannesburg. This is clearly not what is intended as the National Park Board's philosophy is not to run accommodation facilities simply to be hoteliers. Their aim is to persuade people to become interested in the environment.

One of the reasons the Board disposed of the Tsitsikamma Forest National Park and its Storms River Bridge rest-camp in 1987 was because these facilities attracted mainly passing custom, not genuine nature-lovers.

Having said that, however, there are encouraging indications that a considerable number of such casual visitors do book to come back for a longer stay to enjoy a closer look at the national park. This is not surprising given what it has to offer.

The scenery in the park is spectacular and is an attraction in its own right. It is accessible to visitors from both tourist roads and hiking trails while in 1992 the National Parks Board's first four-wheel-drive trail was inaugurated here.

The array of animal life recorded to date within the park's boundaries is quite astonishing. There are 53 species of mammal, 66 reptiles and amphibians (including five species of land tortoise) and 170 species of bird. The presence of the five species of tortoise alone makes the Karoo Park unique because no other proclaimed conservation area in the world is known to have as many species. The park and its environs also accommodate the second-largest known concentration of black eagles after the Matobo Hills area of Zimbabwe.

Then there is the wealth of fossils found here to remind the visitor of what conditions were like 250 million years ago. At that time the Beaufort West area was part of a vast floodplain surrounding an inland sea which covered some two-thirds of

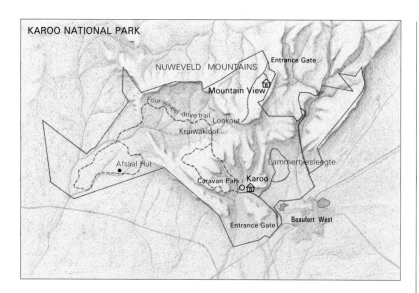

the present land surface of South Africa. Indeed, this part of the Karoo may well have looked much like Botswana's Okavango Delta does today.

Prehistoric reptiles with strange names like *Diictodon*, *Pristerodon* and *Oudenodon* roamed the land. They were not dinosaurs, but rather therapsids – the so-called 'mammal-like reptiles' because they are believed to be direct ancestors of true mammals. This naturally makes them of particular importance in the evolutionary ladder. Enough of them ended up as fossils to give scientists an excellent idea of what the region was like so many millions of years ago, and the Beaufort West area generally is regarded as probably the most important treasure-house in the world for therapsid fossil remains.

The therapsid fossils were formed from the skeletons of animals which died trapped in mud or were drowned by flash floods and then buried in sediments deposited by the flood-waters. Fossilization is only possible if the carcasses are covered quickly by sediments which protect the remains from the disintegrating effects of scavengers, excessive microbial decay, sun, wind and rain. Over many thousands of years the hard parts of the skeleton, for example the bones and teeth, are gradually dissolved to be replaced, bit by bit, by silicate minerals from the sediment deposits around them. Eventually a mineralized replica of the original bones or teeth is produced, often with amazingly fine detail. One of the most famous examples of this process comes from Bavaria where the 140-million-year-old fossil of the first known bird, *Archaeopteryx*, was found in a limestone quarry; its delicate feather structure and reptilian skeleton is clearly shown.

RIGHT: *The flat Karoo plains rise abruptly into the mountainous country which makes up most of the Karoo National Park at present.*

Nocturnal and rarely seen, the aardwolf survives largely on an insectivorous diet and is particularly fond of termites.

The Karoo fossils were enclosed and locked away in the sedimentary rock that had formed around them until, in the course of geological time, the erosive forces of sun, wind and water once again exposed the skeletons to the atmosphere and to the gaze of scientists who can use them to piece together the story of the earth's, and their own, evolution.

The forces at work in this process beggar the imagination because they involve firstly the formation of layers of sedimentary rock and secondly their subsequent removal over a time-scale of hundreds of millions of years. Volcanic activity also played a part in the proceedings: following the period of formation of the fossil-bearing shale and sandstone rock strata, violent volcanic eruptions pumped molten rock from deep within the earth into and through the various rock layers, laying the framework for the Karoo landscape as we now see it.

The molten volcanic rock or magma cooled and solidified to form igneous rock such as dolerite which is very hard and much more resistant to the forces of erosion than the sedimentary rocks around it. Many of the characteristic flat-topped koppies and mountains of the Karoo have been formed because of horizontal dolerite layers and 'caps' which resisted the wind and water erosion that, especially in the southern Karoo, has

removed as much as two kilometres of rock over the last 70 million years. The conical hills and narrow ridges or dykes which are also so typical of the Karoo are often formed around vertical intrusions of dolerite.

All this is made clearer in the Karoo National Park through the Fossil Trail laid out over just 400 metres of veld adjoining the rest-camp. The explanatory brochure for the trail will come as a revelation to the layman because it shows him how to look at the surrounding veld through the eyes of a geologist or a palaeontologist. It transforms the surrounding jumble of rocks at the visitor's feet by pointing out the features that identify the sandbanks and the beds that were formed by living rivers some 250 million years ago. A number of the fossilized remains have been prepared by staff of the South African Museum and are on display in glass cases. The trail was upgraded in 1992 with the addition of larger, more dramatic fossils.

Where the Fossil Trail leaves off, another short but equally educative walk, the 800-metre-long Bossie Trail, takes over. This is designed to remove any impression the visitor may have that Karoo botany consists only of boring grey-green bushes. Markers along the gently rambling trail identify 65 of the common flowers and shrubs which are found in the area, pointing

Cape mountain zebra have been reintroduced to the Karoo National Park as part of the programme to re-establish the species in suitable areas.

out those which are favourite foods of browsing game animals as well as those which are avoided because they are unpalatable or downright poisonous.

One of the commonest of the Karoo Park's 53 mammal species is the rock dassie. The rocky environment of much of the park provides ideal habitat for this agile little herbivore and its abundance is the key to the high population of black eagles in the area. Although black eagles take a wide range of small mammals, dassies are their favourite prey. Work by researcher Rob Davies from the African Raptor Information Centre (A.R.I.C.) has shown that there were 21 breeding pairs in and around the 300-square-kilometre park, 13 pairs actually nesting within its boundaries. Although the dassie population in the park is known to have 'crashed' in 1984, it had recovered to an estimated 13 665 by the 1989/90 season and is possibly capable of rising to 50 000 when conditions are ideal.

An eagle soaring over mountainous country is one of Nature's most evocative sights and the black eagle of the Karoo, displaying its distinctive flight silhouette, provides a glorious counterpoint to the stark Karoo landscape. It is unfortunately a sight no longer common outside the country's national parks and nature reserves because some farmers persistently trap or

Orangethroated longclaws, like many ground species relying on camouflage, are boldly marked beneath but not on their upperparts.

shoot on sight any eagle they encounter, on the grounds that they allegedly prey on small livestock.

Davies's research shows that Karoo farmers have paid a crippling price for their persecution of the black eagle. He points out, however, that it is not a simple case of a healthy black eagle population keeping the dassie population at a permanently low level and thus releasing extra grazing for sheep and goats. As with most ecological situations, it is more complex than that. By removing the black eagle, the dassie's major predator, the farmer is effectively destabilizing the relationship between the dassie and its habitat. Without the eagle the dassie numbers are subject to wild oscillations – irruptions alternating with drastic population crashes. With the eagle, the oscillations are 'damped' and dassie plagues are less frequent and less severe.

If black eagles are removed, the dassies are able to move out of their rocky retreats into the open veld where they compete for grazing with the farmers' small stock. With more food and fewer predators their numbers expand rapidly. When the eagles are present, dassies tend to stay close to their rocky shelters where sentries are posted to keep watch. According to Davies the limit of their range from the rocks is about 12 metres. Within this area the vegetation is heavily grazed and consists of stunted unpalatable bushes while beyond the 12-metre limit grass and palatable bushes flourish.

'Rather than travel another metre to enjoy these rich pickings, the dassies prefer to stay within their limits. It is this abrupt

Armoured ground crickets are most abundant and noticeable in the arid areas of southern Africa. They can be a force to be reckoned with when large numbers periodically invade the veld.

boundary that symbolizes the trade-off that dassies must make between getting enough food and risking getting caught by an eagle', Davies comments. He estimates that a pair of black eagles removes from its breeding territory about a third of the annual potential increase in the dassie population with most of the victims being one-year-old males. At this age the young dassies are driven away from the colonies by the territorial males which means that they have to leave their familiar home grounds and no longer derive any benefit from the colony's sentry system.

Since the park was proclaimed the Parks Board has reintroduced a number of formerly occurring mammals. These include the Cape mountain zebra, red hartebeest and gemsbok, but arguably the most important reintroduction will take place when riverine rabbits are introduced some time in the near future. The riverine rabbit is one of three mammal species listed as 'endangered' in the *South African Red Data Book – Terrestrial Mammals* (1986). Most of its favoured habitat of riverine bush along the seasonally dry river-beds of the Karoo has been degraded over the years by overgrazing, and it is thought that few populations still survive.

Also under consideration is a plan to re-establish a Cape vulture colony in the park at the suggestion of Rob Davies. There is evidence that there was a colony in the park within living memory and it is important to reverse the present downward trend in numbers of this great bird, also a Red Data Book species. Warden Dries Engelbrecht's intention is to use the same system by which the related griffon vulture of Europe has been reintro-

The camponotus ant is easily distinguished from the other species of ants by its yellow abdomen.

duced to the Massif Central mountains in central France. The colony will be started with crippled birds – vultures that have injured themselves on Eskom powerlines, for example, or chicks from an existing colony whose wing bones have been crippled by calcium deficiency. These vultures will be encouraged to breed while healthy young birds will also be brought in and reared in the park in the hope that they will remember this area as 'home' when they are able to fly and to fend for themselves. The project will be run jointly by the National Parks Board and the Vulture Study Group. Rob Davies believes that there is enough potential food within a 70-kilometre radius of the new colony to support a population of 360 Cape vultures.

For the plan to succeed park staff will have to work closely with neighbouring farmers to ensure that the vultures are not shot, or killed by feeding on poisoned carcasses laid out by farmers (which is illegal) to kill problem predators like caracal and the black-backed jackal.

The National Parks Board always tries to be a good neighbour and, in the case of the Karoo Park, this policy has involved it in one of the biggest conservation controversies of recent years – the poisoning of the brown locust. This grasshopper is one of three locust species in central and southern Africa which periodically 'irrupt' in huge swarms and cause much damage to agricultural crops and grazing. The brown locust's permanent home is the central Karoo but when it irrupts it may devastate farmlands as far away as Namibia, Botswana and Zimbabwe.

The last major irruption took place in 1985/6 and was combated by the Department of Agriculture through spraying and dusting with pesticides. There was large-scale use of both BHC and its gamma isomer lindane, both chlorinated hydrocarbon insecticides regarded as deadly environmental poisons. BHC is a mixture of three different isomers of the same chemical and is the most dangerous. It has been banned in many countries, including South Africa, but the Registrar of Pesticides may, at his discretion, permit its use in emergency situations.

The Department of Agriculture seemed unprepared for the 1985/6 outbreak even though locust plagues have been a regular occurrence throughout South African history. It resorted to use of the one-million-kilogram stockpile of BHC, using the Registrar's discretionary power, as part of its 'firefighting' response to the crisis. The trouble with BHC is that it destroys not only the locusts but also the entire spectrum of beneficial or harmless insects so essential to the health of the environment. In addition, animals higher up the food chain which feed on the insects, or which prey on creatures that feed on the insects, build up high levels of the poison within their bodies until it kills them, adversely affects their reproductive capacity or otherwise debilitates them. After the smaller 1986/7 campaign, for example, a dead spotted eagle owl was found with the excep-

tionally high level of 40 milligrams of BHC per kilogram in its body tissues while a dead bat-eared fox (an insect eater) contained 150 milligrams per kilogram.

Such spraying programmes wreak havoc on wildlife in general and in this case the damage could have been avoided in part had the authorities been ready for the outbreak with stocks of less-damaging pesticides. The Parks Board was pressured into spraying locust swarms that entered the Karoo National Park but it did so using the safest pesticide then available, fenitrothion, and in a way that left much of the park untouched. Subsequent minor outbreaks have been dealt with similarly.

What these developments underline is that no national park can exist as an island, separated from its neighbours. Successful conservation practices in the Karoo National Park require close cooperation with neighbouring farmers and organizations such as the Department of Agriculture who have the power to potentially cause damage to the delicate ecosystem of the Karoo.

The Karoo National Park could play a crucial role in saving South Africa's most endangered mammal, the riverine rabbit.

VAALBOS NATIONAL PARK

Although proclaimed in 1986, the Vaalbos National Park is not yet open to the public and the enormous tourist and conservation potential of this national park will only be realized once the facilities, which are currently being planned, are established.

Situated in a transition zone where three ecosystems meet – Karoo, grassland and Kalahari thornveld – Vaalbos rated second only to the biologically richer Maputaland region of northern Natal as an area worthy of conservation in a recent study, using criteria laid down by the World Conservation Union (I.U.C.N.). That result focused the attention of the Parks Board on Vaalbos after the first moves towards having a national park created in the area were made in 1980 by the Barkly West Municipality with the support of the Kimberley Municipality. The park is situated adjacent to the town of Barkly West which lies on the Vaal River about 60 kilometres north-west of Kimberley.

The initial positive assessment of the area's conservation value has been reinforced by further surveys which have shown Vaalbos to be a most useful arm in the project to conserve the endangered black rhinoceros. It will also provide an excellent breeding ground for buffalo that are free of the scourge of foot-and-mouth disease.

The first animals of both species were reintroduced to Vaalbos in 1987 and there is now a breeding population of black rhinoceros, a small group of white rhinoceros and 55 buffalo in the park. The rhinoceros were translocated from Namibia's Etosha National Park and are of the highly endangered Cape subspecies, *Diceros bicornis bicornis*, which is now found only in Etosha and Namibia's Kaokoland and Damaraland regions with two tiny translocated populations in Vaalbos and Augrabies Falls national parks. According to Dr Anthony Hall-Martin, the Vaalbos area is large enough and its habitat is rich enough to make it one of the most important potential sanctuaries in South

The ground squirrel is a commonly seen resident of the dry, western regions of southern Africa.

Africa for the Cape black rhinoceros. (The black rhinoceros conservation problem is dealt with more fully in the chapter on Augrabies Falls National Park.)

The buffalo were brought in from the Addo Elephant National Park and are of great value because they are the only survivors of their species in the Cape Province and furthermore constitute the only natural buffalo population in South Africa free of foot-and-mouth disease.

There is a keen demand for buffalo from southern African game-farmers and nature reserve managers for restocking programmes. Unfortunately this demand cannot be met from the herds in the Kruger National Park because these are carriers of foot-and-mouth disease which is a threat to domestic livestock; in consequence veterinary authorities will not permit them to be moved out of the immediate vicinity of the park. Game-farmers are particularly keen to obtain buffalo as they can earn several thousand US dollars from overseas trophy-hunters for each animal shot. The buffalo is one of the so-called 'Big Five' game animals because of its size, courage, and fearsome reputation.

Vaalbos at present is split into two separate sections with the 'greater' Vaalbos area (formerly the farm Sydney-on-Vaal) covering 18 120 hectares and the Graspan/Holpan region south of Pniel taking in another 4 576 hectares. Before it was taken over as a national park, the greater Vaalbos section was a cattle-ranch on which 2 000 Hereford cattle were successfully run. This means that the land could certainly support at least 2 000 buffalo and selling the surplus animals from a population of this size could result in considerable revenue for the Parks Board.

Other species reintroduced so far to Vaalbos include gemsbok, giraffe, eland, blue wildebeest and the plains (or Burchell's) zebra, bringing the total of large mammal species in the area to 26. The bird checklist presently stands at 189 species. The Board plans to reintroduce roan antelope, tsessebe and hippopotamus. Ultimately, if a minimum park size of 100 000 hectares can be achieved, then the reintroduction of lion, leopard, cheetah and even elephant could be seriously considered.

LEFT: *A herd of red hartebeest about to take flight through grassveld which has turned brown during the long, dry, highveld winter.*

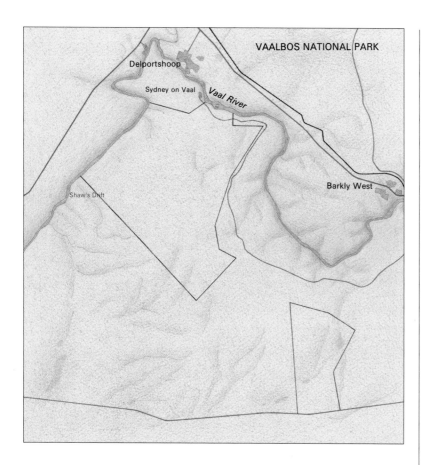

The colourful lilac-breasted roller is at the southern edge of its range at Vaalbos.

However, reaching those long-term goals requires the co-operation of neighbours and other interested parties, a difficulty that has dogged Vaalbos since its inception. The parties concerned question the viability of Vaalbos, not least as a habitat for rhinoceros, and in particular are opposed to the expropriation of privately-owned game ranches to bolster the viability of Vaalbos.

The farm Pniel, owned by the Berlin Mission Society, and the De Beers game reserve, Rooipoort, lie adjacent to and provide a link between the two separate sections of Vaalbos. Rooipoort was bought by De Beers in the 1980's for its diamond potential, but since then has been run as a very successful and well-managed game reserve. In the mid 1980s the Government announced proposals to include, forcibly if necessary, the 42 000 hectare reserve in the Vaalbos Park. De Beers objected to this prescriptive approach as the presence of black rhinoceros and buffalo was incompatible with traditional recreational practices. National park status would also compromise established wildlife utilization enterprises. Consequently the proposals for its incorporation were dropped. More recently, however, discussions between the National Parks Board and De Beers have been resumed in a constructive manner, and provided certain interests of De Beers can be assured, the establishment of a jointly-managed conservation area is possible.

The upper tusks of the warthog are used to dig out bulbs and tubers while the sharp lower tusks are used for self-defence.

The kudu's spiralling horns make it one of Africa's handsomest antelope and it is, fortunately, still common throughout much of southern Africa.

The matter is also further complicated by the fact that some 500 people live as squatters on Pniel with no community infrastructure or facilities. The Parks Board has undertaken to consult with the community leaders on their future and will give preferential employment to Pniel residents in Vaalbos. So far the embryo national park has provided 13 permanent posts for scouts, drivers and artisans and 60 temporary jobs for labourers. The staff complement for a fully-developed park of 100 000 hectares with two rest-camps would be around 15 senior staff, 30 intermediate staff and 150 junior posts. There is currently no enterprise in Barkly West which can offer such employment opportunities. An additional benefit would be the provision of excellent accommodation, with all modern conveniences, in a neatly laid-out staff village.

There the matter lies at present, with discussions still continuing. The development of Vaalbos is being hampered, however, because the best site for a rest-camp is on the banks of the Vaal on ground belonging to Pniel. The Parks Board is reluctant to build on less suitable ground of its own, given that the Pniel land may yet become available in future. Another problem is that, without control of the land on both sides of the Vaal River, the Board cannot reintroduce hippopotamus. At present the na-

The Vaalbos area is large enough, and its habitat rich enough, to make it an important potential sanctuary for the highly endangered Cape black rhinoceros, Diceros bicornis bicornis.

At just over a metre high, the kori bustard is the world's heaviest flying bird.
LEFT: *Extensive stretches of land along the Vaal River will be rehabilitated in the Vaalbos National Park after having been torn apart by mining.*

tional park is fenced off from the Vaal because of the danger of buffalo and rhinoceros leaving the park by crossing the river.

Parks Board Chief Executive Director Dr Robbie Robinson says, 'We have made a proposal but will not force it. If the offer is not accepted then we will start negotiating with the owners with a view to setting up a contractual national park.'

In the meantime, the hard work of setting up a new national park continues. In 1920 the area had a mining population of about 16 000 and much effort has been going into the rehabilitation of sections of ground which had previously been mined for alluvial diamonds. This involves filling in the excavations made by the diggers and then revegetating them.

Other restoration work has involved the breaking up of old high-walled concrete water-troughs used by the cattle and their replacement with lower, contoured drinking-pans that allow both large and small mammals access to the water.

In addition to its natural assets, the Vaalbos area also has considerable cultural and historical attractions. On its one border is the historic diamond-mining town of Sydney-on-Vaal which has retained a number of its original Victorian buildings.

It is to be hoped that the improvements in South Africa's international standing as a result of the recent political developments in the country will aid future negotiations. That being the case, it may not be too long before visitors will be able to view black rhinoceros and buffalo restored to their ancestral haunts in the rolling, golden grasslands and the open savanna veld of the Vaalbos National Park.

KRANSBERG NATIONAL PARK

The Kransberg is a magnificent new national park which is gradually being assembled in the Waterberg mountain range. It has the potential to rival the Kruger National Park's attraction for tourists and its development is one of the top priorities in the Parks Board's overall plans.

Kransberg, which covers 41 000 hectares at present, has yet to be proclaimed a national park because the National Parks Board is still negotiating to acquire the mineral rights over some of the ground. However, depending on the availability of finance, its size could easily be increased to 150 000 hectares and possibly even 300 000 hectares. If this comes about it could also link up to the west with the Atherstone Nature Reserve which is controlled by the Transvaal Provincial Administration and Bophuthatswana's Madikwe Game Reserve to create a park of impressive proportions.

An interesting and important feature of the Kransberg district is that 'the big five' – lion, leopard, elephant, buffalo and black rhinoceros – once occurred naturally in this region. With the exception of the wily leopard which still survives here, all were soon hunted to extinction by settlers, although the last lion in the area was shot as recently as the 1940s. All four species can be reintroduced easily once the park's boundaries are established and fenced, and as the park is just two and a half hours' drive from the major cities of Johannesburg and Pretoria it is clear that it has the capacity to attract large numbers of tourists. The Kruger Park, by comparison, lies a minimum of five hours' drive from these centres.

The re-establishment of 'the big five' as well as other large mammals at Kransberg will help the Parks Board to reduce the growing pressure of tourists on the Kruger National Park. There are two reasons for this. Firstly, the majority of tourists and

The Kransberg's elevation allows many species to grow which would not survive in the surrounding bushveld, such as this Crossandra greenstockii.

visitors to national parks go there to see big game; the scenery, birdlife and other attractions are secondary. At this stage in South Africa only the Kruger Park can offer 'the big five' along with the spectacle of large herds of animals such as wildebeest and impala. Secondly, being so much closer to Johannesburg and Pretoria, the Kransberg Park will be a more convenient national park for both overseas as well as local tourists to visit. The same reasoning lies behind the recent establishment of Bophuthatswana's 75 000-hectare Madikwe Game Reserve which is currently being stocked with the big game appropriate to the region.

But there is much more to Kransberg than simply being a convenient locality for seeing big game. For those naturalists with a broader interest in the environment a strong case can be made for the claim that the Kransberg is actually more interesting than the Kruger Park. The dominant Sour Bushveld veldtype of the Kransberg area will support most of the large mammal species found in the Kruger Park, but the montane region adds an ecosystem which is missing from the latter. The Sour Bushveld vegetation of the Kransberg has, in fact, many elements in common with the bushveld found around the Punda Maria and Pretoriuskop sections of the Kruger Park.

Visitors will also find the Kransberg Park's climate to be milder than that of the Kruger Park as it is on average 1 000 metres higher in altitude. Summer temperatures are cooler and the area does not suffer from the stifling humidity so typical of the eastern Lowveld in normal rainy seasons.

The lowest plains in the Kransberg Park lie at an elevation of 1 020 metres, but the mountain peaks reach up to 2 088 metres and introduce a whole new range of plants in this more temperate habitat. In the montane grassland, for example, at least 10 species of ground orchid have been identified so far; there are also four species of protea, and other elements of the fynbos vegetation found in the southern Cape and the higher parts of the Transvaal Drakensberg such as ericas. Then there is the splendid Waterberg cycad (*Encephalartos eugenemaraisii*) which

LEFT: *Hardly in familiar surroundings, a yellowwood sapling grows at 2 000 metres on the roof of the Kransberg National Park.*

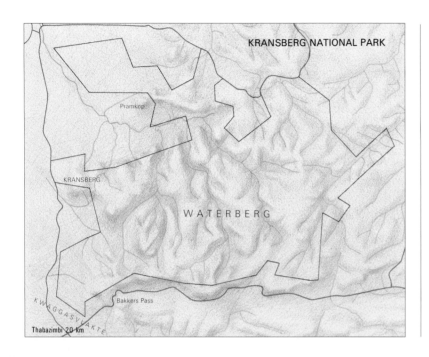

is endemic to the Waterberg massif. Also present are trees typical of the Afromontane forests of the southern Cape such as the real yellowwood, assegai and stinkwood. To date 393 plant species of 93 families have been recorded in the park.

Among the notable features of the upland parts of Kransberg National Park are a number of vital vleis and wetlands which absorb the high rainfall in the mountains and then release a regular flow of water into the rivers which drain the region. These wetlands are known to hydrologists and ecologists as 'sponges', for obvious reasons, and any damage inflicted on them by, for example, overgrazing and compaction of the earth by cattle, will result in heavy water run-off following rain which can cause erosion. In addition the rivers fed by the sponges will stop flowing in the dry season if the sponge is no longer able to act as a reservoir to supply them with retained water.

As the park is situated in the transitional zone between the dry western and the better-watered eastern regions of South Africa, this again contributes to the diversity of plants and

Rain falling on these rugged slopes and ridges eventually winds up in the Limpopo River and flows to the Indian Ocean.

The civet is another of Africa's numerous carnivores which is seldom seen because of its nocturnal habits.

animals in the area. The effect will become more marked in the park when it is extended north and west on to the low-lying plains surrounding the Waterberg.

Piet van Staden, the first warden at Kransberg, only took up his post in 1988 and, although the compilation of an inventory of the park's plants, birds, mammals and reptiles is a priority, it has had to take second place to the more urgent demands of transforming former cattle-ranching country into a national park. These include taking down stock-fences that divided up what was formerly cattle range, putting up game-proof boundary fences, constructing roads and finding the best site for the proposed rest-camp. Then there is the delicate task of establishing diplomatic relations with neighbours used to grazing their cattle and other livestock in areas which are now part of the national park but not yet properly fenced off.

Despite all these diversions and the fact that Van Staden's particular interest is botany, his bird checklist already exceeds 200 and, judging by the diversity in similar areas around Kransberg, there seems little reason why the total should not eventually reach 400 within the park's boundaries. The montane habitat brings in a number of species typical of the mountainous areas of the Drakensberg and which are not found in the Kruger

RIGHT: *A sight for sore eyes in an area frequently wracked with drought. The Matlabas River drains the Kransberg and flows into the Limpopo River.*

Park; these include buff-streaked chat, Cape rock thrush, malachite sunbird and Gurney's sugarbird. The last two species are present because of the proteas on which they feed.

The Waterberg is also home to the largest remaining breeding colony of the Cape vulture which is listed as 'vulnerable' in *The South African Red Data Book – Birds*. The latest survey has shown around 800 active breeding nests in the colony which is about half the total for the Transvaal. The colony does not lie within the present boundaries of the park but is situated on cliffs just outside it. Negotiations are under way with the farmers owning this ground with a view to having the breeding cliffs incorporated in the national park.

The prime motives for setting up the Kransberg Park were to conserve areas of the Waterberg range together with the Sour

Bushveld veld-type. The initial development was intended to be a cooperative venture between private companies and the State, with the State buying a central core of ground totalling 15 741 hectares and the private companies contributing the rest as a contractual national park. The State expropriated the central core of land in March 1988 – a move which caused much ill-feeling amongst the local farmers and which is only now abating according to Van Staden. The land intended to be added as a contractual national park, however, never materialized. In the end, the farmers owning this land sold it directly to the State without compulsion or expropriation.

Almost the entire national park at present consists of mountainous ground although there is a low-lying area in the north where three tributaries of the Limpopo River – the Matlabas, the

Despite its name, the black mamba is never really black. The inside of the mouth, however, is uniform black, a characteristic that differentiates it from most other snakes.

The Kransberg mountains are home to the largest breeding colony of Cape vulture in the Transvaal. This colony is not, as yet, part of the national park.

Proteas, nurtured by the high rainfall and cool temperatures, blossom at a height of 2 000 metres in the Kransberg National Park.

Mamba and the Sterkstroom – debouch from the Waterberg range. The Parks Board hopes to enlarge the park here.

The list of large mammals which formerly occurred around Kransberg and which will eventually be reintroduced makes fascinating reading. In addition to 'the big five' already mentioned, they include giraffe, roan antelope, sable antelope, cheetah, wild dog, spotted hyena, Burchell's zebra, oribi and grey rhebuck. So far only the reedbuck has been reintroduced to join the species still found naturally in the park: these include kudu, bushbuck, mountain reedbuck, klipspringer, warthog, bushpig, aardvark, brown hyena, leopard and caracal. Should the park's boundaries eventually be extended far enough westwards, the possibility exists that both springbok and gemsbok – animals typical of the arid west – could also be brought back.

So far about 30 species of reptile have been identified in the park. These include three threatened species – the rock leguaan, the water leguaan and the python. There are numerous snake species, mostly harmless with one noteworthy exception being Africa's deadliest snake, the black mamba.

Further development of the Kransberg Park and the speed at which it takes place will depend on the availability of funds from the Parks Board. Van Staden says the opening up of the park to tourists is one of his main concerns because of the benefits this will bring to the nearby community of Thabazimbi.

The first visitors to experience this new park will be of a more adventurous nature as plans are underway to develop either a four-wheel drive trail or a hiking trail. A picnic-site for day visitors is also planned for the near future to cater particularly for people from nearby Thabazimbi. This area was a popular picnicking area for residents before access was restricted.

It has not yet been decided when a full-scale rest-camp will be constructed at Kransberg, but it is unlikely there will be such a facility before 1997, by which time certain of the roads within the park will have been upgraded to allow access to normal vehicles. Until then, the only way of experiencing this remarkably diverse new national park will be the hard way – in a four-wheel drive or by donning hiking boots and rucksack and stepping off on a scenically magnificent trail.

KRUGER NATIONAL PARK

The Kruger Park is one of just three reserves in South Africa where wild dogs are found.

The Kruger National Park was born and raised in the teeth of opposition and controversy and now, nearly a century after the first steps were taken to inaugurate it, controversy has again risen over the future of this park, one of the world's greatest natural wildlife sanctuaries.

Developments over the last decade have brought home to the park's custodians the harsh message that the fate of this two-million-hectare segment of wild Africa is inextricably linked to developments in the rest of southern Africa. It simply cannot be run as if it were an island, a separate entity. The ecological management of the Kruger Park is increasingly affected by the industrial, social, economic and political events taking place outside its boundaries. The issue of conservation is very much part of the turbulent political change as the 'New South Africa' is created, and as Mozambique, which adjoins the Kruger Park along the entire length of its eastern boundary, strives for an end to years of internal conflict.

Many of the key problems faced by the research and field staff originate outside the park's borders and the solutions often lie outside the conventional approach to nature conservation.

One of the most pressing of these is the looming shortage of water which threatens to dry up most of the Kruger's seven major rivers for significant periods each year. Until relatively recently five of these rivers had a strong perennial flow. The reasons for the reduced flow lie in the demands being placed on available water supplies in the Lowveld by agricultural and industrial development and by the booming population explosion taking place on the park's boundaries. In addition, water still flowing through the rivers into the park is increasingly subject to pollution by all these major consumers.

The Kruger Park has also been affected by the political upheavals that have racked southern Africa since 1974. The park has not remained unscathed, for there have been massive incursions of thousands of Mozambicans attempting to find refuge in South Africa from the bitter and ruinous hostilities in their own country. Some of the practical problems park managers have had to face as a result have been numerous uncontrolled veld fires started by the refugees and the re-emergence of reports of man-eating lions and leopards which have preyed on the unfortunate fugitives. No man-eating lions had been recorded in the area since about 1900.

Civil war in Mozambique has also resulted in large-scale poaching in the park as automatic weapons are readily available. This new threat has forced park field staff to undergo paramilitary training in order to be able to defend both their wildlife charges and themselves.

Commercial-scale poaching in the park reflects the harsh social reality of South Africa as a Third World country with a rapidly expanding rural population that is outgrowing its available resources. The visitor to the Kruger National Park who drives in through the overgrazed and eroded subsistence communal farming areas will readily appreciate the magnitude of the problem when he reaches the lush, conserved ground that starts immediately at the park's boundary fences.

In a new political order the park may be threatened as impoverished farmers and new voters could question the rationale behind the park's existence and demand that the land be distributed. A.N.C. leader Nelson Mandela stated on British television in October 1991 that his party would not lay claim to land inside the Kruger Park and that the park was an 'established fact whether we like or not how it came about'.

The Parks Board is now under pressure to justify the park's existence to its immediate neighbours by providing them with opportunities to participate in the benefits which it offers. The people have to be shown that utilization of this vast area as a national park is of greater long-term value than subsistence agriculture and they have to be convinced somehow that the conservation ethic is not simply a moral imperative but the key to man's survival. The Kruger Park is a major employer (there are 3 400 people in its work force) and a crucial component of

LEFT: *A marula tree spreads its bare branches over a winter landscape. The tree bears a heavy crop of fruit in summer which is eagerly sought after by animals ranging from baboon to elephant.*

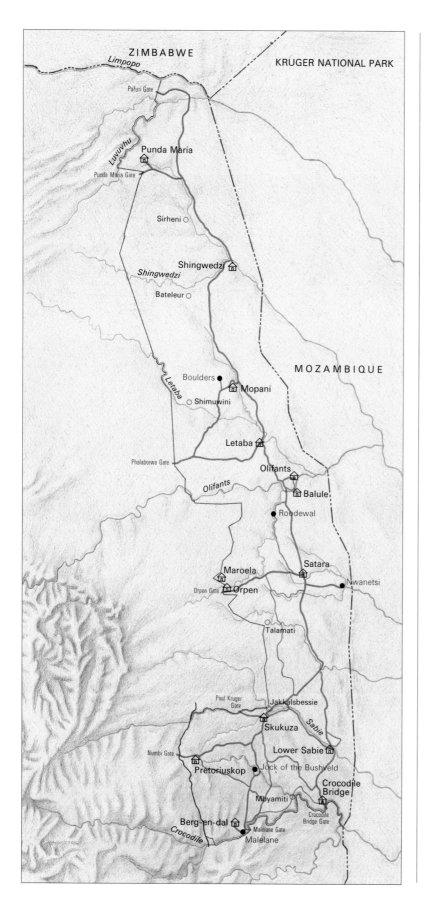

the economy of the Lowveld in that it attracts over 700 000 tourists to the area every year. To its credit the Board has already set up an intensive environmental education and awareness programme aimed at schoolchildren and community leaders from the neighbouring black communities.

Recent political developments in both South Africa and Mozambique have also allowed the Parks Board to play an active role in international conservation diplomacy. Plans for an international peace park are now well advanced. The extension of the Kruger Park's eastern boundary into neighbouring Mozambique could perhaps double the park's present size. In addition a creative agreement has been reached with the private game reserves along its western boundary, which allows the intervening fences to be taken down. Should a proposed link northwards into Zimbabwe's Gonarezhou National Park ever materialize, the end result will be the world's greatest wildlife sanctuary.

So today the Kruger National Park faces a somewhat uncertain future with, on the one hand, tremendous potential and, on the other, a number of deeply worrying problems. Nevertheless it is a situation which would not have surprised or unduly disconcerted the man who set up this greatest of African wildlife sanctuaries from virtually nothing and ran it for the first 44 years of its existence – Lieutenant-Colonel James Stevenson-Hamilton.

He lived with similar – and worse – situations from the earliest days of the Sabie Game Reserve, forerunner of the Kruger Park, to which he was appointed head ranger in 1902. These were summed up in his book *South African Eden* in the delightful dedication, 'To Cinderella who became a princess.' Anyone reading this book will realize what a remarkable man he was, with a vision decades ahead of his time. Apart from the practical difficulties he experienced, Stevenson-Hamilton also faced the daunting challenge of changing an attitude strongly held by the general public – that of the individual's inalienable right to hunt anything, anywhere, without control.

However, before delving too deeply into Stevenson-Hamilton's contribution, we must go right back to the origins of the Kruger National Park which lie with that remarkable man after whom the park is named, Stephanus Johannes Paulus ('Oom Paul') Kruger.

He was a sixth-generation Afrikaner of trekboer stock who became President of the South African Republic in 1883. If he is remembered at all internationally, it is as the man who defied the might of the British Empire and fought the Second Anglo-Boer War against Britain from 1899 to 1902. After Pretoria was occupied by the British, ill-health prevented him from taking

RIGHT: *In order to prevent damage to the vegetation through overpopulation, elephant numbers in the Kruger Park have to be kept down by culling.*

Bull hippopotamuses guard their territories fiercely and, with immense jaws and lethal teeth, ward off intruders with a threatening 'yawn'.

LEFT: *Hippopotamuses take to the water and safety. Drought can take its toll on these animals when rivers such as the Luvuvhu and Limpopo dry up completely.*

part in the guerrilla phase of the war and he went into voluntary exile in Holland in 1900, eventually to die in Switzerland in 1904.

Biographers dwell on his negative personality characteristics and paint him as a dour, gruff, coarse, stubborn and unbending man, yet it was he who first challenged the prevailing viewpoint of the day – that wildlife was there to be hunted – and proposed that land be set aside as a sanctuary for game. It was tantamount to heresy at the time for a nation which believed it had an absolute right to hunt and to use all its available land for agriculture. Paul Kruger, looking back on what had happened to the game populations of the Transvaal since the Voortrekkers trekked there in the 1830s, had the vision to realize that something had to be done if anything of the wild heritage the Afrikaners had taken for granted was to survive.

His initial proposals in 1884 to the Volksraad, the parliament of the South African Republic, were rejected but they led eventually to the proclamation in 1894 of Africa's first game reserve, the Pongola Game Reserve. The reserve covered more than 15 600 hectares of ground along the north bank of the Phongolo River in the wedge of Transvaal between Swaziland and Natal. Unfortunately this reserve was deproclaimed in 1921 following years of neglect, although a small Transvaal provincial nature reserve was re-established here on 27 February 1980.

Kruger persevered with his campaign to establish a reserve in the eastern Lowveld until, on 26 March 1898, he was at last

Nyala are found only in the far north of the Kruger Park because they cannot tolerate cold weather. They are often seen along the banks of the Luvuvhu River.

able to sign the proclamation setting up the 'Gouvernement Wildtuin' (Government Game Reserve) between the Sabie and Crocodile rivers. Later to be known as the Sabie Game Reserve (and to become the Kruger National Park in 1926), it did not get off to a good start because a year later the Anglo-Boer War erupted. In the three years of strife that followed, the rules governing the management of game reserves became largely irrelevant. The Sabie Reserve came under the control of a free-booting soldier called Ludwig Steinacker who commanded an irregular British military unit known as Steinacker's Horse. Stevenson-Hamilton had to deal with Steinacker immediately after the Anglo-Boer War when the former was appointed to the

post of head ranger of the Sabie Game Reserve on a 'temporary' two-year secondment from his regiment.

According to Stevenson-Hamilton, 'Steinacker's Horse had experienced all the advantages, with few of the disagreeable drawbacks, of being on active service.'

It was a turn-of-the-century utopia for those accustomed to a freebooting life style in the bush, but one which Stevenson-Hamilton quickly abolished in accordance with his instructions from Sir Godfrey Lagden – the man who appointed him to the post – to make himself 'generally disagreable'.

Stevenson-Hamilton's position was tenuous in the extreme. He comments that his mandate was 'vague, general and ver-bally conveyed to me by people who were nearly as ignorant of the conditions prevailing as I was myself'. Further, as he sub-sequently realized, 'So far, much that I had done had been *ultra vires*; there were no special regulations to support me, and I possessed no judicial or other power, such as that of arrest, under the civil law.'

From such small and uncertain beginnings emerged one of the world's finest wildlife sanctuaries. Stevenson-Hamilton made as big an impression on the local black population as he did on Steinacker's Horse. He removed squatters from within the Sabie Reserve's boundaries and ruthlessly stamped out poaching. The people's response was to nickname him 'Sku-kuza' – 'he who turns everything upside down'. One of his workers articulated the general feeling, as Stevenson-Hamilton recounted in *South African Eden*: 'As one of my attendants re-marked to somebody, "Never have I travelled with such a white man; when he saw a zebra standing so close that I could have hit it with a stone, he only looked at it. Truly, he is quite mad!"'

With Steinacker's Horse and the squatters removed, Stevenson-Hamilton's monumental task had just begun. In the next two years he increased the total area of the park to over 36 000 square kilometres – almost eight times the 4 600 square kilometres of the original Sabie Reserve.

However, Stevenson-Hamilton was now about to enter the most bitter phase in the development of his 'Cinderella'. He attempted to persuade the authorities that the area should be declared a national park. Almost immediately he found himself in conflict with vested interests in the form of the land com-panies which had other ideas for the ground as the country's economy started to boom in the wake of the declaration of the Union of South Africa in 1910. The companies which had granted the original game protection concessions between the

RIGHT: *Giraffe bulls fight standing side by side and swing their heads and necks at each other. These fights are seldom fatal and usually end when the animals are too exhausted to continue.*

The white rhinoceros derives its name from the Dutch word 'wydmond' meaning wide mouth. The broad, square mouth is adapted specifically for grazing.

Sabie and the Olifants up until 1912 did not want to extend them, while the Selati railway was extended north from Sabie Bridge to link up with Soekmekaar north-west of Tzaneen. The pressure built up as farmers demanded grazing rights during winter in the Pretoriuskop area. They eventually succeeded despite Stevenson-Hamilton's strenuous opposition. Thousands of sheep entered the reserve in 1912 and it was not until Pretoriuskop was at last brought into the Kruger National Park in 1926 that the graziers were finally excluded.

Then came the First World War and Stevenson-Hamilton rejoined his regiment, virtually giving the powers ranged against the reserve a free hand. After much agitation in certain quarters a Commission was appointed by the Union Government in 1916 to look into the game reserve issue. The Commission was generally favourable to the achievements of Stevenson-Hamilton and his staff and also recommended that the two reserves, Sabie and Shingwedzi, should be proclaimed a national park. Nothing was actually done about this at the time but, as Stevenson-Hamilton said, 'the stage was definitely set'.

Further difficulties soon arose. By 1921 coal-mining companies were prospecting deposits around Komatipoort while the Selati railway line had proved a financial flop and the company running it was asking for farming rights along the line to improve its economic situation.

The nadir was reached in 1923 when the Secretary of the Department of Lands announced at a meeting held in Pretoria that his Department wanted to abolish the entire Sabie Game Reserve for subdivision into farms. But, just when all seemed doomed, events at last started to swing Stevenson-Hamilton's way as wider support for a national park began to emerge amongst the public and certain important Government personalities.

A change of government at this stage brought into the equation the third key man in the formation of the Kruger National Park – the new Minister of Lands, Piet Grobler. Fortunately this grandnephew of Paul Kruger came out wholeheartedly on the side of the national park concept, countering the onslaught of the farmers and would-be land-grabbers. In April 1926 Grobler and his colleagues came to an agreement with the companies with land inside the Sabie Reserve, in which they gave up their ownership either in exchange for Government-owned land outside the reserve or for a cash settlement. In addition the Govern-

A crocodile swallows its prey of fish while a saddlebilled stork and a grey heron hover, ready to snatch it if it is dropped.

ment allocated the wedge of land between the Olifants and Letaba rivers to the reserve, thus joining the Sabie and Shingwedzi reserves and leaving the almost-to-be-declared national park with approximately 1,9 million hectares of ground and a shape not much different from what it is today.

Grobler introduced the National Parks Act in Parliament on 31 May 1926, ensured that it passed without opposition and announced that the Sabie Game Reserve would change its name to the Kruger National Park. A long and bitter struggle was over and the park has never looked back.

What these three men – Kruger, Stevenson-Hamilton and Grobler – had created was a great tract of wilderness some 350 kilometres long from north to south and 90 kilometres across at its widest point. It contains one of the largest assemblages of life on earth with 300 tree, 49 fish, 33 amphibian, 114 reptile, 507 bird and 147 mammal species – to say nothing of the myriad forms of invertebrates, many of them still unknown to science. Dominating this great array is a population of 7 500 elephants which includes some of the last great tuskers of Africa, mighty, battleship-grey bulls carrying more than 45 kilograms of ivory a side. This is one of the very few elephant populations in Africa

that is growing, to the extent that it has to be culled annually to prevent overpopulation. It is worth remembering that when Stevenson-Hamilton did his initial tour around his new reserve in 1902, he was sorely disappointed at how few elephants he found and in 1912 it was estimated that there were only 25 left. However, these were supplemented by large numbers which subsequently came into the sanctuary of the reserve from neighbouring Mozambique.

The boundaries of the Kruger Park have remained more or less unchanged since 1926 when its proclamation as a national park made its status official and secure. There are now three major developments for the park which may transform it from being just one of the 'top ten' of the world's national parks to being – arguably at least – the best of them all.

The first of these is an initiative started by the Mozambican Government and supported by the Southern African Nature Foundation which holds out the possibility of creating a national park adjoining the Kruger Park incorporating between four and six million hectares of Mozambican territory. The area under consideration has been depopulated by the bitter civil war waged in Mozambique between the ruling Frelimo party and

the Renamo opposition since the early 1980s. The World Bank has subsequently agreed to provide funds for a detailed management and project study but the prerequisite for any development is a lasting peace in Mozambique. Once that has been achieved Parks Board Chief Executive Director Dr Robbie Robinson says the World Bank has made it clear it will fully support the establishment of the new national park.

The project will be run by the Mozambicans with the South African National Parks Board acting as consultants in the short term and providing as much support as it can. Once the Parks Board is satisfied that conditions in Mozambique have improved to the point where the survival of large game animals can be reasonably assured, then large numbers of animals of different species will be sold or donated from the Kruger Park to the new park. This means, for example, that up to 400 elephants could be transferred to Mozambique every year instead of being culled in the Kruger Park as they are at present. Depending on how successful the new national park will be, Robinson indicates that eventually the Kruger Park's fence along the Mozambican border could be removed. In addition the combined area would be run by a joint management committee.

The second major development in the Kruger Park's 1990s transformation is the possibility of coming to an agreement with the Zimbabwean Government whereby the combined park would also link northwards to join up with the Gonarezhou National Park in Zimbabwe. This would create by far the world's largest national park spread across three countries.

Consummation of this vision is still some way off and has, it appears, been overtaken by the third major development in the Kruger Park's imaginative programme to destroy man-made barriers to wildlife movement. This is the long-awaited removal of the fences on the park's western boundary after decades of bitter wrangling between the Parks Board and the privately owned nature reserves in the Klaserie, Timbavati, Umbabaat and Sabi-Sand areas. The existing private game reserves such as Mala Mala and Londolozi will continue operating as before in their own areas, taking out clients in open vehicles and at night. The Parks Board will manage the game in the entire area.

When the western fences eventually come down, the game will have access to a new water source in the form of the Sand River and the hope is that the traditional east-west migration of large ungulate species such as zebra and wildebeest will be re-established. This is by no means certain, however, as the game herds have adjusted to new patterns of movement since the fences were erected.

LEFT: *A redbilled oxpecker can be seen perched on the back of this buffalo from which it forages its main diet of ticks.*

The Kruger Park is the main stronghold in the Transvaal of South Africa's largest eagle, the martial eagle. It is seen here feeding on a young kori bustard.

Water has become a crucial issue for the Kruger National Park because of the twin pressures of consumption and pollution placed on the seven major rivers which flow from west to east through this diverse park. Ironically, the park has had to fence off its northern boundary from a sixth river, the Limpopo, because this is an international boundary and there is no game reserve on the Zimbabwean bank. This could be rectified by the proposed agreements with Mozambique and Zimbabwe.

The rivers help split the park into some of its distinctive landscapes of which there are four main ones reflecting the drop in annual rainfall from south to north and the broad geological split of the park into granite structures on the western half and basalt on the eastern. This division is of necessity a very broad one because altogether 35 different landscapes have been identified in the Kruger Park.

North of the Olifants River the vegetation of the park is dominated by the distinctive mopane tree (*Colophospermum mopane*) and this region is in fact the southern end of the vast mopane forests that stretch northwards into the dry and lower-lying areas of central Africa. North towards the Limpopo the flat and generally low mopane savanna and scrub is increasingly broken up by the bulk of the broad-girthed baobabs towering above the lesser trees and with the silhouette of their twisted branches creating one of the most striking of African landscapes.

Cheetah and hyena face each other in a confrontation which the more powerfully built hyena are sure to win.
RIGHT: *It is unusual to see lion feeding on rhinoceros as these animals are more than a match for a lion and therefore rarely attacked.*

This panorama is what many visitors see in their mind's eye when they think of Africa and it is very different from the regions south of the Olifants where the vegetation is considerably more diverse. The south-east of the park is largely open savanna with typical trees being the knob thorn (*Acacia nigrescens*), marula (*Sclerocarya birrea*) and leadwood (*Combretum imberbe*), while in the south-western districts is a mixed combretum veld with thorny thickets and dense stands of various species of trees. The fourth major type of vegetation is the magnificent riparian forest which borders all the major rivers. For the visitor, the most easily seen stretches are along the Luvuvhu in the north of the park.

For decades park authorities have taken for granted the water that flowed through its seven rivers – the Luvuvhu, Olifants, Crocodile, Sabie, Letaba, Limpopo and Shingwedzi – with flows affected only by the periodic droughts that grip southern Africa. However, in the mid-1960s Kruger Park managers realized that the droughts were not entirely of nature's making. They believed that the unprecedented agricultural, exotic forestry and industrial development of the Lowveld regions was seriously affecting what could be called the 'water balance' of the park and jeopardizing its primary conservation mission to maintain genetic diversity as well as essential ecological processes and life-support systems.

They discovered their rivers were carrying increased loads of sediment and silt caused by erosion as the result of poor land use practices. They were also experiencing periodic bouts of pollution from industries discharging waste into these rivers,

particularly from the phosphate mines at Phalaborwa which falls within the Olifants River catchment area. This river was shown to carry very high levels of sulphates and fluorides. In addition, huge fish kills occurred after discharges of silt from the Phalaborwa Barrage resulted in a 'mudstream' that smothered life in the river. The Crocodile River on the park's southern boundary carries sewage effluent from the surrounding rural areas as well as high salinities and pesticide residues from the intensive irrigation farming operations taking place on its south bank and in its upper catchment area.

Reduced volumes of water caused by man's overexploitation of the rivers created all kinds of other problems, such as concentrating hippo populations in certain areas and thus causing overgrazing damage to adjacent riverine vegetation. To make matters worse, the reduced quantities of water flowing in the rivers make them less able to deal with the periodic spills of pollutants because these cannot be diluted quickly enough. In the event, the Kruger Park authorities realized they had to fight for water rights of their own, and subsequently laid a claim with the Department of Water Affairs for the national park to be considered a legitimate consumer of water along with the other traditional consumers – agriculture, forestry, industry and the domestic householder.

The River Research Programme was started in 1986/87 to determine just how much water and of what quality is needed to maintain a river ecosystem, to ensure the survival of fish and invertebrate species as well as hippopotamus and crocodile, and to prevent elements of the riverine forest from dying off?

The relationships are numerous and complex. After the die-off of fish in the Olifants River following the first silt spill, Pel's fishing owl, a rare raptor that lives mostly on fish, became rarer in the area and five species of fish disappeared from the river. The Parks Board is working on an agreement that Phalaborwa's sluice-gates be opened only in summer during high-water flow conditions that will effectively dilute the effect of the silt and other pollutants downstream.

The message being expounded by Parks Board researchers is that rivers are longitudinal ecosystems that cannot be divided up into isolated sections. The rivers draining eastwards through the Kruger Park are integral parts of the park's life and should be treated as watercourses of great national and international importance. The River Research Programme's results are significant and far-reaching and will undoubtedly be scrutinized with interest by park and reserve authorities in regions far removed from the Transvaal Lowveld.

The results of the River Research Programme will also have an impact on the park's ability to maintain the populations of spectacular mammal species such as the lion and leopard, and the bigger herbivores such as elephant, buffalo and rhinoceros. These animals are the major attractions for the majority of the nearly 700 000 visitors who come to the park each year.

The Kruger National Park is home not just to these mammals but also to a wide range of lesser-known and sometimes more vulnerable species. These include the African wild dog and the cheetah and, in particular, a number of the larger species of predatory birds such as bateleur and martial eagles which are becoming increasingly rare outside protected game reserves.

The approach adopted by park managers is to interfere as little as possible in the natural workings of the park and there is indeed a rule of thumb in conservation to the effect that the larger the reserve the less one has to interfere. Although the Kruger Park is a very large reserve, the fact that it is artificially enclosed by fences means that management has to intervene on occasion to maintain what researchers deem to be a suitable 'balance'. One obvious example in the Kruger Park is provided by the elephant: if its numbers are allowed to expand without control, it has the capacity to destroy not only its own habitat, but also that of many of its fellow creatures.

Ground hornbills are normally seen in small groups of between two and four birds, carefully searching the ground for their prey which consists of insects, lizards and even large snakes.

Leopards are rarely seen, being a shy and mainly nocturnal species.

Understanding how the system works, developing management plans, observing the results and, if necessary, 'fine-tuning' the plans are all functions carried out by the park's research staff. Of particular importance is the work being done on the relationships between the various predator species and their prey. Recent work on wild dogs and cheetahs has been the most intriguing for the visitor because, to compile the data needed for their research work, the park's scientists have relied heavily on information gleaned from their thousands of visitors.

The wild dog has been the focus of particular attention because it is the continent's most endangered large carnivore. It has been estimated that there may only be between 2 000 and 5 000 left in sub-Saharan Africa and, of these, about 360 are to be found in the Kruger Park.

The relatively low numbers of wild dogs in the Kruger Park, however, pose a paradox which is neatly contained in the question asked by Parks Board researcher Dr Gus Mills who is currently studying the wild dog: 'If the wild dog is such an efficient hunter, which we know it is, then why is it so scarce in the Kruger Park which contains in excess of 125 000 impala, its favourite prey?' From that question was born his research programme on the wild dogs of the park.

The wild dog research project required detailed statistical information on the numbers of dogs in the park as well as comprehensive biological data on pack structure and behaviour. To accomplish the latter Mills intended to dart members of various known packs in the southern section of the park between the Sabie and Crocodile rivers and equip them with radio collars so that their movements could be monitored on the ground and also from space by satellite. Firstly, however, he required an accurate census of the park's dog population as well as a means of identifying pack members. Mills and his colleague Anthony Maddock chose the unusual course of enlisting the help of the public, reasoning that the park's excellent road network and large numbers of tourists would ensure regular wild dog sightings.

Together with the Endangered Wildlife Trust they therefore organized a competition asking tourists to take photographs of any dogs they saw and send them to their Skukuza office with details of time and place. Each wild dog has a unique coat pattern of black, white, brown and yellow blotches and the two researchers hoped to build up 'identikit' pictures of every individual dog in the park. The response was overwhelming and over five thousand pictures were submitted during the

A Euphorbia rowlandii *spreads its fleshy arms tipped with red flowers against a background of lichen-covered boulders.*

18-month campaign. By combining this photographic record with tracking data from the second phase of the project, Mills was able to refine his population figures and identify the various packs of dogs and their home territories.

One of the findings of the research project was that for some unexplained reason there is an unusually high mortality rate amongst adult wild dogs, around 36 per cent per year. Yet despite this the population remains stable. Dogs caught and subjected to veterinary examination are invariably found to be very healthy and so disease can probably be ruled out as the cause of the deaths. There are other possible reasons: man with his poaching snares, for example, seems to have a negligible effect, or predation by lions, which Mills believes could be a more important factor than has hitherto been believed. Once, for example, he witnessed three lions ambush a pack of wild dogs and kill seven of its pups. If this were the case, it poses the question of what, if anything, the park authorities could do to increase wild dog numbers.

The answer will lie in the management strategies. Mills rules out the direct action that was common in the past such as artificially holding down lion numbers to promote wild dog.

Many people still have a negative attitude towards the dogs. They are viewed as vicious and indiscriminate killers although research by scientists like Mills and the Frames and Von Lawicks in East Africa, has shown them to have a close and highly developed communal family system and that they kill only to eat.

Mills and co-worker Tony Boland are now working on a project on the park's small cheetah population using the same technique of a competition to get visitors to send in their photographs. Like the wild dog with its individual coat pattern, each cheetah has a unique spot pattern. Although good, the response to the competition could not compare with the runaway success of the wild dog project and it appears that cheetah are simply not so easy to photograph. Wild dogs usually ignore vehicles and this can result in spectacular sightings when the packs sit on or near the game-viewing roads.

Lion numbers in the park are stable and, due to the provision of water and consequent stabilization of prey, probably higher than they have ever been. The lion's status today for visitors is exactly as commented on by Stevenson-Hamilton writing on conditions in the mid-1920s. He noted then that initially it was policy to control their numbers but, when the national park was declared 'the lions, instead of being considered a menace ... suddenly, having acquired immense popularity with the sightseeing public, became [the park's] ... greatest asset'.

Park authorities have imposed particularly strict measures to protect visitors in comparison with controls laid down in other game reserves, particularly in countries like Botswana and Zimbabwe. All of the park's rest-camps, including the bush-camps and the trail base-camps where the emphasis is on getting as close to nature as possible, are securely fenced following three incidents in the 1960s at Olifants Camp when tourists sleeping out in the open on the verandas of their huts during hot nights were bitten by hyenas.

The only animals which park managers believe they have to control by culling or capturing are elephant, buffalo and hippopotamus because of the serious damage they can inflict on the environment when their numbers become too great. The elephant population is kept at approximately 7 500 and buffalo numbers are maintained at around 25 000. Meat from the culled animals is processed at an abattoir near Skukuza and is sold to tourists, used in the restaurants or sold to staff at special rates as well as to the nearby communities from which the Kruger Park has recruited its 3 400 workers.

The ivory was a valuable source of revenue until the trade was banned in January 1990 by C.I.T.E.S. (the Convention on International Trade in Endangered Species of Wild Fauna and Flora). This left the Parks Board with stocks of unsold ivory worth around R6-million at a recent assessment. Such an amount of money is sorely needed to fund the conservation effort in South Africa. The ivory controversy is problematic. South Africa, like Botswana and Zimbabwe, has succeeded so well in conserving its elephant herds that they are expanding and require culling. In contrast the herds of most of the countries of West, Central and East Africa have been devastated by poaching. The ban on the international ivory trade implemented at the motivation of these countries, and supported by most First World countries, has stopped southern African conservation agencies from realizing an important and legitimate, albeit problematic, source of income.

Chief Executive Director Dr Robbie Robinson upholds the right of South Africa to sell ivory from its well-conserved herds but concedes this is a delicate situation because it has become such an emotive issue. He advocates a step-by-step approach to the resolution of the problem instead of outright confrontation, with the southern African countries winning firstly the right to sell the hides from culled elephants and then secondly the right to sell their ivory. He believes that the solution to the impasse lies in the drawing up of an agreement between the African countries that hold the elephant populations; it will not come from the overseas nations that consume the ivory. The reason for urgency on the sale of hides is that, unlike ivory which lasts for ever, hides deteriorate and become worthless if not cured into leather within 18 months.

The culling programmes exclude the big tuskers for which the Kruger Park now has an international reputation. In the late 1970s the park became famous for its 'Magnificent Seven' – the seven largest bulls in the park. Between 1981 and 1984, however, all these bulls died either from natural causes or because of poaching, except for one, Joào, who broke off both his majestic curved tusks about 20 centimetres from the lip line. The reason for this is not known but elephant expert Dr Anthony Hall-Martin, now Executive Director: Southern Parks, speculated it was in a fight of herculean proportions.

The ivory from all these superb bulls has been collected by the Parks Board and a number of sets of tusks will form part of an information centre focusing on elephants which was set up at Letaba Camp in 1992.

With the demise of the Magnificent Seven, however, more of the park's enormous bulls came into the limelight. One of these, Phelwana, had to be destroyed in 1988 after he had been wounded by poachers outside the park. His tusks proved to be a record for the park at 75 and 55 kilograms respectively. Nevertheless, after his death there were still another four exceptional

Kudu are mainly browsers, feeding on leaves, fruit and pods. Here a kudu bull browses on a mopane tree, the dominant vegetation over much of the northern section of the Kruger Park.

A baby pangolin hitches a ride on its mother's back as they cross a game-viewing road.

bulls in the park with tusks weighing well over 45 kilograms each, while at least another 10 had been identified, with tusks approaching the 45-kilogram mark, all located around Tshokwane and along the Salitjie road. The Kruger Park is now the best locality in the whole of Africa for a chance of seeing a truly great tusker.

Poaching remains an ever-present threat to the existence of these magnificent animals, as well as to the game populations in the park in general. So far, however, the area has not experienced the total poaching onslaught that has overwhelmed game reserves in other African countries but, as Executive Director: Kruger National Park Dr Salomon Joubert comments, 'It would be foolish to think we are immune. It is a serious concern given the plans the Parks Board has for building up its black rhinoceros population from the present 200 to the 3 000 that ecologists believe the park can support.'

Some 200 elephant bulls are known to have been poached by Mozambicans between 1981 and 1984 and another 27 were killed between January and October 1991. Kloppers points out that while this was serious, park authorities managed to apprehend most of the poachers responsible. He adds that the park's well-trained and highly motivated staff, supplemented when necessary by South African Defence Force personnel, can deal with any poaching threat to the region's wildlife.

However, Dr Salomon Joubert brings the final answer on poaching back to the general state of the country: 'Good nature conservation depends on social, political and economic stability. If that collapses we are in trouble.'

The very success of the Kruger National Park in attracting visitors has brought an unanticipated problem in its wake: for now not only must the park authorities ensure that the carrying capacity of its larger mammals is not exceeded, it must see to it that the flood of human visitors does not reach the level where the wilderness experience the Parks Board has tried so hard to promote will be compromised. This is a very real concern as anyone who visits the park during peak occupancy periods such as long weekends or school holidays will quickly appreciate.

By the end of 1991 the Parks Board had completed a 10-year programme to increase its available beds from 3 000 to 4 000 by increasing accommodation at some existing rest-camps and by constructing a number of entirely new camps. The number of camping-sites remains the same. Dr Joubert comments that the park will now go through a consolidation phase of perhaps five to eight years before the Board will consider providing any further accommodation.

'During peak periods I go around the park to get a "feel" for the atmosphere because we have to maintain the Kruger's wilderness ambience at all costs. When we have forfeited that, the Kruger will have lost its soul. My personal feeling is that we have managed to retain the feeling of wilderness so far, but we are now approaching the danger point of having too many people', he says.

Joubert says the Parks Board's philosophy on accommodation is to 'keep it simple, keep it wild' as well as 'affordable and accessible'; it is not the intention to provide facilities above three-star level. However, although the luxury accommodation and five-star treatment is, by and large, left to the private game reserves, it does appear that some of the facilities being provided in the new accommodation being built in the Kruger and other national parks must be considered to be of luxury standard.

What surprises many visitors is the park's extensive network of tarmac roads. One consequence of this is that the Parks Board has to run its own traffic control department which issues fines to those motorists tempted to break the speed limits. Johan Kloppers comments that the tarred roads are far more 'environment friendly' than the traditional gravel roads found in most game reserves and they are also cheaper to maintain.

The Kruger National Park has many moods and many faces: they vary from season to season as well as over longer periods, from the lush, green times of plentiful rain to the lean years of drought when the earth is scorched by the sun and the painfully thin antelope seek listlessly for scarce shade and scarcer food.

Each season has its advantages and its disadvantages. The winter months are dry and, although much cooler than the summer months, the Lowveld is still pleasantly warm with

RIGHT: *Ever alert, impala drink warily against a backdrop of riverine forest along the Luvuvhu River.*

Of all the different insect species in the Kruger Park, butterflies are without doubt the most beautiful.

maximum daily temperatures in the mid-20s Celsius and cool but frost-free nights. This should be compared with the freezing conditions of a typical highveld winter night or the wet, windy and chilly winter season of the Cape. The beginning of winter in the Lowveld marks the beginning of the rutting season for the park's commonest antelope, the impala. The territorial rams herd their compliant and graceful ewes, tirelessly chasing off the non-territorial bachelors.

In other respects, the pace of life in the park tends to be slower in winter although the birds of prey choose this time to breed in order to have their young leave the nest at the time of plenty when the rains come. For those visitors who wish to see large concentrations of big game the best time is towards the end of winter when the antelope and other game animals tend to be concentrated around the remaining waterholes waiting for the rains to arrive. Great herds of buffalo gather around the water, kicking up clouds of dust through which the late afternoon light filters in pastel shades of gold and orange.

Summers in the park can be cruelly hot, with the temperatures occasionally soaring above 40 degrees Celsius while the humidity builds up to oppressive levels. Yet for those who can tolerate the heat with or without the aid of air-conditioned cars

and huts, this is the time of plenty when the park is alive with migrant birds, the resident birds don their bright breeding plumages and the game animals have their young.

Human conversation outside the restaurants in camps like Skukuza and Lower Sabie is sometimes almost drowned out by the noisy, rasping calls of breeding masked and lesser masked weavers. The yellow-and-black males chatter harshly as they build their round basket nests and then display from them to attract a mate. The female usually rejects the first few nests, forcing the male to tear his perfectly woven edifice apart and start the whole construction process again. In the middle of this commotion, 'Nature, red in tooth and claw' takes its course for those visitors patient enough to look for drama amongst the park's multitude of lesser creatures.

If they are lucky they may see a fierce-eyed Gabar goshawk raid the colony, creating pandemonium as it extracts a chick with its beak, or perhaps they might catch sight of a metallic-green and white diederik cuckoo provoking the weavers to mob him while his drabber mate sneaks in to lay her egg in one of the weavers' clutches. The young cuckoo usually hatches before the baby weavers and ejects the eggs or defenceless babies of its host from the nest so that only it will be fed by the parents.

The rain, when it comes after the months of drought, can be a highly evocative experience. Blue-black thunderclouds build up over the waiting mopane and baobab trees wilting under the oppressive heat and humidity while lightning flickers in the distance and the thunder mutters growing threats about the storm to come. The clouds grow larger and the wind rises until, in a shattering climax, the skies open and rain pours down to refresh the dry earth and the unforgettable smell of the bush after the first rains rises to delight the senses of the visitor.

In a matter of days the landscape is transformed as the bush metamorphoses from its winter shades of dusty khaki to vibrant mopane green, and for two months or so the baobab trees look quite odd as their leaves sprout, in contrast to their usual stark and twisted contours.

At any time of year the visitor can expect to experience the sights and events that typify the African bush experience: the swooshing sound of vultures as they drop out of the sky to land before bounding into the scrum of birds squabbling over the remains of a kill, while a pride of lions lies replete in the shade nearby; a marabou stork outlined against a swollen, red African sunset; a huge solitary elephant bull crossing the road in front of a vehicle, supremely confident of his right of way.

For those determined to get the most out of their bush trips, perhaps the best way to do it is to book for one of the seven wilderness trails operating in the Kruger Park. These are the Bushman, Wolhuter, Napi, Metsi-Metsi, Sweni, Olifants and Nyala trails. A maximum number of eight people can go on each

A spotted hyena and its young warm themselves in the early morning sun.

trail at any one time and they take two days and three nights to complete. Trailists are taken into the bush by experienced, armed rangers who are very willing to share their knowledge of the bush with the hikers.

It's one thing to get close to big game in a vehicle, it's another to find oneself in close proximity on foot, particularly because it can happen so unexpectedly. What was open bush a second ago is now filled with one ton of huffing, puffing, irascible black rhinoceros or four tons of majestic, striding bull elephant. A stalk on foot behind a party of elephant bulls in bushveld is an experience for all five traditional senses and the sixth sense one perhaps never knew one had. They are tracked by following the huge 50-centimetre diameter footprints and are usually heard long before they are seen – elephant tummy rumbles and throat growls reverberate through the bush accompanied by the pistol-like cracks of branches being broken off. A trail group's approach is determined by the wind and the ranger's assessment of the situation.

Buffalo may look something like cattle from the safety of a car, but to the hiker in open ground, computing distances and sprinting times to the nearest climbable tree, they take on an infinitely more menacing air. They raise their noses and glower

at intruders, fixing them with an unwavering stare from below the huge bosses of their horns. Often they are curious and will take a few steps towards the party, apparently unable to believe what they are seeing. It can be a terrifying but exhilarating feeling for the trailist and at times like this unquestioning obedience to the ranger's commands is the order of the day.

It is, unfortunately, extremely difficult to get on these trails – bookings have to be made months in advance and even then it is not easy to obtain a place. However the Parks Board intends to develop more trails in the near future.

A criticism often levelled at the Parks Board is that it has not provided enough facilities to allow visitors to leave their cars. One of the great attractions of Zimbabwe's Hwange National Park and the Natal Parks Board reserves is that hides are provided where visitors can sit and watch the ebb and flow of wildlife around waterholes. The Parks Board does intend to remedy this situation.

It can only make one's experience of the Kruger that much better and, if the ambitious expansion plans for this park materialize, whole new vistas will be opened up in Mozambique for the hordes of regular visitors for whom the Kruger Park is their favourite bush experience.

GOLDEN GATE HIGHLANDS NATIONAL PARK

Visitors used to the style of accommodation and facilities in other South African national parks are in for a surprise when they first visit the Golden Gate Highlands National Park, for this park – at first sight – appears to be nothing more than a beautifully situated holiday-resort.

The main 'camp', Brandwag, is a hotel complete with conference facilities and ladies' bar, and all accommodation units in the main building and its associated motel-like chalets have their own television sets and telephones. Other facilities include bowls, tennis, table-tennis, snooker and horse-riding, while during the holiday seasons the staff organize a daily activity programme for visitors which includes canoeing, abseiling and guided hikes.

The park does offer the more traditional hutted accommodation in the Glen Reenen Rest-camp. It is interesting to note, however, that changes in some of the other national parks may result in them developing as sophisticated a range of facilities as those provided at Brandwag as the Parks Board adjusts to meet people's changing demands.

The reason for the different approach to visitors' needs adopted at Golden Gate Park lies in the nature of the park itself. It was proclaimed as South Africa's first scenic national park in 1963 to protect the magnificent landscape surrounding the spectacular sandstone cliffs in that part of the north-eastern Orange Free State which adjoins the mountain kingdom of Lesotho. It gets its name from the towering 100-metre-high golden-yellow sandstone cliffs, the Golden Gate, that flank the valley of the Little Caledon River at the western entrance to the park.

Warden Corrie Pieterse maintains that many younger people prefer his park to the Kruger Park. 'There's so much more to do here because of all the recreational activities that we offer. We want to become known as a national park which has special attractions for young people.'

A female stonechat keeps watch for insects from a vantage point above the surrounding grasslands.

LEFT: *Golden winter grasslands roll across the eastern Orange Free State to the mountains of the Golden Gate Highlands National Park.*

His comments are borne out by remarks from information officers at a number of other national parks who are often approached by younger visitors after one or two days' stay with the inevitable question, 'Isn't there anything else to do here?' It would seem that the average young adult of today needs more to encourage relaxation than merely looking at all that beautiful, but apparently boring, scenery and wildlife. How, and whether, the National Parks Board should cater for them is the subject of considerable debate.

Apart from the obvious motive of making profits, there is another reason why the Parks Board wants to attract young visitors here: environmental education. The natural attributes of the park make it an outstanding outdoor classroom for studying the earth sciences of geography and geology, but also provide opportunities for the practical communication of the environmental conservation message.

Education has therefore been accorded high priority at the park, and the Wilgenhof Youth Hostel is operated specially for visiting school groups. The hostel accommodates up to 80 children at a time and can cope with up to 45 school groups annually. The programme is so popular it is booked out a year in advance. The park is also the venue for the annual National Youth Symposium to which the winners of various national environmental competitions organized by bodies such as the Department of Environmental Affairs and Wildlife Society are invited.

Although the primary educational aim of the Parks Board is to spread its environmental message, the schoolchildren who probably gain most from a visit to the park are the students of geography. For displayed before them at Golden Gate is much of what is in their text-books. The park shows excellent examples of the various kinds of weathering and erosion as well as the effects of the ice age that gripped southern Africa between 10 000 and 1,8 million years ago.

The park lies between the altitudes of 1 892 and 2 837 metres and, geologically, is formed from the upper part of the Karoo Sequence. The various layers were laid down between 230 and

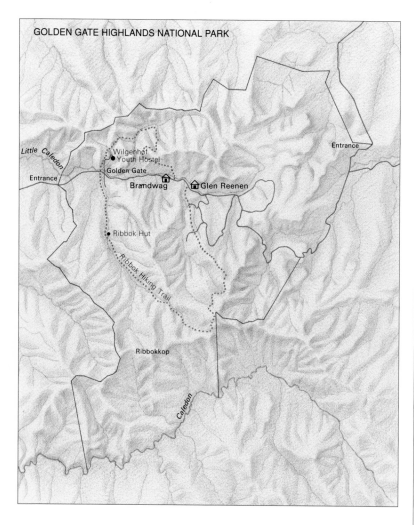

GOLDEN GATE HIGHLANDS NATIONAL PARK

190 million years ago through sedimentation by both wind and water over a period during which the climate became progressively drier until the area turned into arid desert. The layman's attention might tend to wander when technical terms like Karoo Sequence and sedimentation are used but, at the Golden Gate Park, understanding the geology of the area is made very much simpler by the fact that the various features are clearly identified by their distinctive colours.

The bottom layer is the approximately 23-metre-thick Molteno Formation which consists of coarse sandstone with layers of khaki-coloured mudstone. It is only visible in the eastern section of the park and is best seen at the drift over the Klerkspruit on the road to Kestell.

The next layer is the approximately 150-metre-thick Elliot Formation which is a reddish-brown mudstone. It rates a high priority in the Golden Gate Park's management policy for two reasons – it contains dinosaur fossils and it poses a serious erosion problem. An important discovery was made in 1978 when a clutch of six dinosaur eggs was found, in three of which the developing embryo dinosaurs were clearly visible.

Two important features of the Elliot mudstone are that it erodes very easily and vegetation does not grow readily on it. It is also prone to 'slumping', causing mudslides, if exposed to heavy rain. Consequently, according to research officer Gideon Groenewald, park managers go to considerable trouble to ensure that the formation is not needlessly exposed. Careful planning is therefore required in the construction of roads and footpaths and in the design of veld-burning programmes. If a footpath crosses the Elliot Formation, its surface has to be sealed otherwise serious erosion is inevitable.

The next layer in the Karoo Sequence is the Clarens Formation which, for the visitor, is the most noticeable because the spectacular yellow sandstone cliffs like the Golden Gate and the Brandwag buttress opposite the Brandwag Rest-camp are part of this formation. It is a very fine sandstone which was deposited by wind under desert conditions and it varies in thickness between 140 and 160 metres. A distinctive feature of many of the cliffs is the black vertical striping which is formed by lichen growing on the rocks where rain-water seeps down the faces.

The sedimentation processes that built up the Karoo rock layers came to an abrupt end about 190 million years ago when tremendous volcanic eruptions shook the earth and vast quantities of basalt lava were forced up from the bowels of the earth to cover the surface. The basalt makes up the topmost geological layer in the region and is known as the Drakensberg Formation. It caps all the mountains in the park and, on the highest peak, Ribbokkop, is 600 metres thick.

That same volcanic action is also responsible for the shape of some of the park's most spectacular scenery because not all the lava flowed over the earth's surface. Some of it was forced sideways into horizontal cracks in the existing rocks and, for example, where the lava was forced into Clarens sandstone, the great heat baked the sandstone into a hard quartzite. The quartzite was considerably more resistant to erosion by wind and water than the surrounding sandstone, and eventually created the landscape features now known as the Mushroom Rocks and the Cathedral Cave.

The Mushroom Rocks were formed by the erosion of the sandstone under the quartzite plate at the top of the cliff, leaving the plate overhanging like a veranda. A similar process resulted in the spectacular Cathedral Cave, a 60-metre-deep chamber carved out of solid sandstone but with a quartzite roof. The river which hollowed out this cave plunges through a hole in the roof to create a dramatic waterfall during the rainy season.

As with the Kruger Park, the Golden Gate Highlands National Park may be poised to grow in unexpected directions in the next few years. Following its expansion in 1988 from 6 241 hectares to 11 630 hectares, the park now borders both the independent nation of Lesotho and the self-governing state of

QwaQwa. If all goes well, it is possible that there could be a merger of the Golden Gate Park with some 20 000 hectares of ground belonging to QwaQwa that has been declared a natural area by the QwaQwa Government. Several Parks Board officials sit on the board of directors of QwaQwa's Tourism and Nature Conservation Corporation, the body responsible for the ground.

If this merger comes into being, there is the further possibility of joining up the QwaQwa nature area with similar stretches of land in Lesotho, and then extending across to the Royal Natal National Park in Natal. Despite the name, this is not a national park in terms of the National Parks Act and is controlled by the Natal Parks Board. If this ambitious plan were to be realized, a combined Drakensberg National Park of impressive dimensions would have been created. To achieve such an outcome, however, would involve overcoming formidable obstacles in international politics, as well as reaching an agreement between rival conservation organizations – perhaps an even more daunting task! Relationships between the National Parks Board and the various provincial nature conservation bodies in South Africa have frequently been strained.

In the short-term, Golden Gate Park's most pressing problems concern the correct management of its game animals and the long-standing feud over the provincial road that runs through the park from the Orange Free State to QwaQwa. Although numerous applications have been made to the Orange Free State authorities, the National Parks Board has so far been unable to obtain permission to put up gates on this road. As a result, the Parks Board has no control over traffic entering the Golden Gate Highlands National Park.

Vehicles speed past both rest-camps on their way through the park, while herders and their dogs occasionally pass through with their flocks of sheep and herds of cattle. Nothing can be done to prevent it provided they keep strictly to the provincial road. There is a never-ending littering problem along the road and veld fires are occasionally started.

The issue of managing the park's game populations is not as easy as it might seem. Managers have to weigh the tourist spectacle of abundant large game against the surprisingly limited carrying capacity of the park's vegetation. Gideon Groenewald points out that 13 per cent of the park's present area

Thousands of years of erosion have carved the sandstone rocks of the Golden Gate Highlands National Park into many fascinating landscapes.

of 11 640 hectares is bare rock, and much of the rest of it supports sour grassveld which most game animals find unpalatable. Censuses over a number of years have in fact revealed that the game herds use a mere 4 000 hectares of the total area. There is therefore a real risk of overgrazing which, in turn, can easily lead to a serious soil erosion problem on exposed valley slopes. The black wildebeest population in the park has been reduced from 330 to 180 because of this danger.

According to Groenewald, the only large herbivores in the park that would normally spend all their lives in mountainous areas are the grey rhebok and the mountain reedbuck. There are 400 of each species in the park. All the others such as Burchell's zebra, black wildebeest, blesbok and eland are essentially plains animals which would spend only a limited period in the mountains when rainfall and grazing conditions were suitable.

Groenewald stresses that control over game numbers is vital to the protection of the park's ecosystems, particularly in the case of species like springbok which are prolific breeders. Surplus animals are either culled or sold to game-farmers. Future culling and capture of game will take place mainly in the mountainous regions of the park to encourage the antelope to move out of the high ground and down to the newly-acquired plains.

Among the most distinctive plants at Golden Gate Park are two species of alien trees which are allowed to flourish contrary to the normally very strict Parks Board rules. These are the

Golden Gate is home to one breeding pair of bearded vultures which in southern Africa are restricted to the mountain kingdom of Lesotho, and surrounding areas such as the Drakensberg.
LEFT: *Willow trees line a lake below the sandstone formations. High rainfall in the surrounding mountains makes this a well-watered region.*

These overhanging formations, the Mushroom Rocks, were formed by rock layers which were more resistant to erosion than the surrounding sandstone.

weeping willows and Lombardy poplars along the Little Caledon River that runs alongside the main road through the park. They have been permitted to remain because they have been there so long that they are an accepted part of the scenery of this region of the Orange Free State – a policy ruling that has its opponents. Other aliens such as wattle and bluegum, however, are being systematically eradicated.

Indigenous trees are scarce in this region with the most common being the oldwood or 'ouhout' (*Leucosidea serica*) which is a member of the rose family and grows in dense stands in the valley bottoms. Another tree likely to be encountered is the distinctive mountain cabbage tree or 'bergkiepersol' (*Cussonia paniculata*) spreading its floppy crown on the mountain slopes.

The acquisition of the flatter terrain in 1988 allowed the National Parks Board to reintroduce a spectacular rare lizard – the sungazer or giant girdled lizard – which is endemic to the Orange Free State and very limited adjoining sections of the Transvaal and Natal. It is classed as 'vulnerable' in the *South African Red Data Book – Reptiles and Amphibians*. This is the only national park with appropriate habitat for the sungazer.

As with many mountain areas in South Africa the birdlife of the Golden Gate Park is not particularly rich and only some 140 species have been recorded so far, compared with the 400-odd species one might expect to find in an equivalent area of bushveld. However, some of the birds present are unusual and relatively rare species, well worth looking for: they include, for example, the bald ibis, bearded vulture and Cape vulture.

The bald ibis is endemic to mountainous areas of southern Africa and occurs only in Lesotho, the eastern Orange Free State, eastern Transvaal, Swaziland and the mountainous western districts of Natal. There is a breeding colony of about 28 pairs of this bird in the Golden Gate Park. Once considered endangered, the bald ibis has remained stable in numbers for the past 20 years with a total population of around 3 000 birds, although ornithologists are still monitoring the situation carefully. This striking bird with its conspicuous red crown is a swift and powerful flier and, although it is normally silent, its hoarse calls may be heard echoing around the breeding cliffs in the park.

The bearded vulture's flight silhouette is unmistakable, with its immense wingspan reaching nearly two-and-a-half metres and its relatively long wedge-shaped tail. This magnificent bird's range in southern Africa is nowadays restricted to the mountains of Lesotho and adjacent parts of South Africa and a sighting is a prize for any bird-watcher. Its speciality is shattering bones by dropping them from heights of between 50 and 150 metres onto a flat rock; it then glides down to eat both the bone fragments and the richly nutritious marrow. The bearded vulture has an enormous home range – up to 4 000 square kilometres – and it is not surprising therefore, that only one pair is found in the Golden Gate area. These birds are considerably easier to see at the Giant's Castle Game Reserve where the Natal Parks Board runs a 'vulture restaurant' for them during the winter months. The first 'vulture restaurant' run by the National Parks Board is to be erected at Golden Gate. The East Rand region of the National Parks Board's Honorary Rangers' has initiated the raising of funds for this project which came into effect in the first quarter of 1993.

The area also supports a population of around 40 Cape vultures which roost in the park but have not yet formed a breeding colony. This cliff-nesting vulture is listed as 'vulnerable' in the *South African Red Data Book – Birds*. Its future depends on the protection of the remaining breeding- and roosting-sites and on the campaign to persuade farmers that vultures should be fostered as useful members of the agricultural community. A minority still erroneously believes that vultures are stock-killers while others are careless in their use of poisons for destroying such genuine stock-killers as jackal and caracal – this use of poison is in fact illegal. Being highly efficient scavengers, the vultures are often the first to reach a poisoned carcass and an entire flock may be lost as a result.

A series of short day trails, as well as the 30-kilometre-long Rhebok Trail which takes two days to complete, threads its way through this mountain world. If the proposed link with the contiguous natural area in QwaQwa should come to fruition, hikers can expect this network to be extended into the Maluti Mountains. Hiking is undoubtedly the best way to experience the superb scenic beauty of this country. Nevertheless it is well to remember that climatic conditions can change swiftly in mountainous regions – snowfalls are common in winter and the weather can become bitterly cold at any time of year.

Such sudden changes in the weather are as much an integral part of the visitor's experience of this mountainous national park as the dramatic landscapes. Should the grand scheme of extending the Golden Gate Park into QwaQwa and Lesotho ever be achieved, then a vast mountain hinterland will be conserved and opened up for exploration by hikers and other visitors.

A heavy snowfall blankets the sandstone mountains. Snowfalls are a regular feature of the Golden Gate area during the winter months.

MOUNTAIN ZEBRA NATIONAL PARK

Having succeeded in its objective of bringing the formerly endangered Cape mountain zebra back from the brink of extinction, the Mountain Zebra National Park is now facing a battle of a different kind: it is striving to become a financially viable asset in the conservation portfolio run by the National Parks Board.

As the Board responds to South Africa's changing political climate and the need to become financially independent, the future development of this particular park could make it a fascinating business case study, in addition to a conservation success story, because of the challenges it faces in its undertaking to attract more visitors.

There are numerous credits on the Mountain Zebra National Park's balance sheet. It is set in the magnificent, mountainous Karoo country near the town of Cradock in the eastern Cape, an area of particular historic and literary significance in South Africa because this is where Olive Schreiner lived, taught and based her famous novel, *The Story of an African Farm*. In addition to the Cape mountain zebra, the park has herds of several other species of large game such as eland, red hartebeest, kudu and black wildebeest. At least 200 bird species have been identified so far. There is also a three-day, two-night hiking trail which rates as one of the finest in South Africa.

The essential attraction of the Karoo is that it has remained virtually unchanged since European settlement began 340 years ago. This is in direct contrast to most other areas of South Africa which have been subjected to intense economic development and huge increases in human population. The herds of game that roamed the vast plains of the Karoo have long gone and the vegetation has been altered and degraded because of man's impact – or that of his sheep and goats – but the Karoo still remains much as it was, an immense, empty land of wide-open spaces broken by the distinctive flat-topped koppies.

LEFT: *Granite mountains rise above the surrounding Karoo in the Mountain Zebra National Park.*

South African hedgehogs are generally slow moving but can show a surprising turn of speed by rising high on their long legs.

Above all, however, the Karoo is a place of silence, in marked contrast to the noise and bustle which otherwise govern the lives of 'civilized' man. If one stops one's car on any empty stretch of road in the Karoo, all that can be heard is the sound of the wind in the telephone wires and perhaps a bird call or two.

Despite the fact that it has so much to offer, the Mountain Zebra Park during 1991 had an average room occupancy of only 60 per cent and an average bed occupancy of 52 per cent, while a mere 557 people walked the Mountain Zebra Trail giving it a use rating of just 13 per cent. By comparison, the Karoo National Park has already achieved a room occupancy of 70 per cent even though its rest-camp was only completed two years ago. The new rest-camp facilities at Mountain Zebra Park have been in use for more than 10 years.

Mountain Zebra Park warden Etienne Fourie attributes this to lack of advertising and the fact that, unlike the Karoo National Park which lies on the main route between Johannesburg and Cape Town, his park is situated off the beaten track. The National Parks Board, however, has recently started to advertise specific parks which should help to encourage more visitors to the park.

There has also been an unfortunate trend over the past few years, following the large accommodation price increases by the National Parks Board, for even fewer people to visit the Mountain Zebra Park. Even though these visitors are spending more, it is not a welcome situation from the business point of view, and it deviates from the Parks Board's stated aim to make its facilities affordable to a wide spectrum of the population.

Looking at ways and means of making more money out of tourism is focusing Fourie's attention on the essential role of a national park and what changes could be made. 'What you are looking at is essentially greater commercialization of a national park along resort lines because that seems to be what visitors want. We would then have to supply facilities like swimming pools and children's playgrounds but there's always the question of how far we should go. Do we, for example, start catering

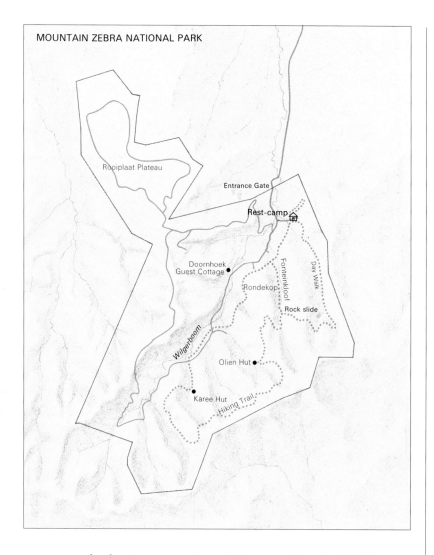

MOUNTAIN ZEBRA NATIONAL PARK

Rooiplaat Plateau

Entrance Gate

Rest-camp

Doornhoek Guest Cottage

Fonteinkloof

Day Walk

Rondekop

Rock slide

Wilgerboom

Olien Hut

Karee Hut

Hiking Trail

for functions or parties which we could easily do? Whatever happens, certain principles are non-negotiable. The national park is an asset for all the people of South Africa and I cannot allow any activity which will degrade or endanger this asset', he comments.

Such concerns are vastly different from the issue which led to the founding of this national park in 1937 – the urgent need to conserve the few remaining Cape mountain zebra in order to prevent this close relative of Hartmann's mountain zebra from following the blue antelope of the southern Cape into the oblivion of extinction. The success of the park in doing this is a matter of record.

However, like the Bontebok National Park which was also set up with the sole purpose of preserving a single mammalian subspecies, the early results in the Mountain Zebra National Park were disappointing and local farmers played a key role in keeping the park going.

There were five stallions and a single mare present on the 1 712 hectare farm Babylonstoren when it was declared a na-

tional park, but by 1946 the mare and three of the stallions were dead. In 1950 five stallions and six mares were donated by a neighbouring farmer and the augmented herd had increased to 25 by 1964. In that year, five of the adjoining farms were purchased, bringing the park's area to its present size of 6 536 hectares. As a bonus, however, one of the farms had a healthy population of 30 zebra and from this combined total of 55 animals the population has not looked back.

In addition to the more than 200 zebra in the Mountain Zebra Park there are, at the time of writing, a further 80 animals in the Karoo National Park, 11 in the Bontebok National Park and 18 in the Zuurberg National Park, giving a total of around 345 animals under Parks Board protection. There are also over 180 Cape mountain zebra in six of the nature reserves controlled by the Cape Provincial Administration and at least another 180 in local authority reserves and private game reserves. From a low point of about 100 animals in 1950, the Cape mountain zebra recovered to over 400 in 1984 and the world population in 1992 stood at just over 700 animals. Over 660 of these are derived from Mountain Zebra National Park stock – a tribute indeed to the dedicated conservationists of the National Parks Board.

The Board's future policy is to continue to relocate zebra to appropriate habitat throughout the animal's former range, on the principle that if the eggs are spread amongst enough baskets the subspecies will be protected from catastrophic events such as disease wiping out one of the major populations. National parks and provincial nature reserves would be accorded preference, followed by local authority reserves, private nature reserves and private game farms.

Fourie believes fully in the eventual commercialization of the Cape mountain zebra through trophy-hunting on private farms. 'The fact is that if conservation pays it stays, and the main commercial value of a Cape mountain zebra is its skin as a hunting trophy. Game is a renewable resource and one of the dangerous side-effects of action by the Green movement, such as bans on animal skins, is that they can remove the value of the animal and, in the process, a major incentive for the private sector to conserve it.'

One particular aspect of the Mountain Zebra Park that will immediately strike the visitor is the concentration of game on the area known as the Rooiplaat Plateau around which curves one of the game-viewing roads. At any given moment about 85 per cent of the large herbivores in the park will be found here, the reason being that Rooiplaat consists of sweet grassveld which the grazers far prefer to the sour grasses and Karoo scrub covering most of the rest of the park. This concentration of game on one relatively small area, which occurs because the park is not divided into 'camps' and the animals can roam freely where they choose, has raised fears of overgrazing that could cause long-term damage to the veld. Given the pattern of recurrent droughts which characterizes the Karoo, this is certainly a distinct and unwelcome possibility.

Rooiplaat, which is covered by rich grassland in years of good rainfall, takes on the grey appearance of typical Karoo shrubland in drought periods. Before farmers fenced off the Karoo, the larger game animals would have trekked away from the region in dry periods to find better grazing; as that is no longer possible, they continue to feed on Rooiplaat eating the most palatable of the shrubs and leaving the least palatable to multiply.

Parks Board researcher Peter Novellie believes such small, well-stocked and well-fenced reserves probably lack the resilience to withstand a succession of droughts; this means plant and animal species could be lost as a result. He says the Mountain Zebra Park has coped well so far but queries whether it could survive the chance succession of severe drought years that may come once in a century. 'Is it undergoing a gradual but relentless decline in the relative abundance of palatable food plants coupled with slow erosion of irreplaceable topsoil, decade by decade? At present there is no evidence of such potentially disastrous trends but continued vigilance over a long period is required before we can be fully confident', he says.

At present the Mountain Zebra Park covers an area which is mostly mountainous country. Although the Parks Board would like to enlarge it, firstly to achieve a better balance between mountains and flatter areas, and secondly to increase the car-

Mountain reedbuck set against a background of rolling Karoo veld and conical koppies.

rying capacity for the mountain zebra, nothing has happened since the acquisition of the five extra farms in 1964. Fourie says the reasons for this are twofold: land acquisition for other parks has taken priority, and local farmers have been asking exorbitant prices for land around the Mountain Zebra Park.

He maintains that land in this part of the Karoo should sell for around R300 per hectare, but the owner of one farm the Parks Board is interested in acquiring has been asking R1 300 per hectare. 'There is just no way we can ever pay that kind of inflated price', he comments.

Another bone of contention between the Parks Board and the local farmers in the area concerns the alleged activities of the dominant predator in the Mountain Zebra Park, the caracal. This large reddish-coloured cat with the distinctive tasselled ears is rarely seen by visitors. The caracal's extraordinary caution and nocturnal habits have allowed it to survive in many stock-farming areas despite the most strenuous of attempts by farmers to eradicate it.

Adult male caracals weigh up to 20 kilograms and they can prey on animals as large as mountain reedbuck. Research on the caracal in the Mountain Zebra Park and the surrounding areas by Lucius Moolman, now warden at Addo, has shown that its main prey species in the park is the rock dassie which made up 53 per cent of its diet during the study period, followed by the mountain reedbuck which accounts for another 11 per cent.

Outside the park the dassie is still the most important prey item but only accounts for 30 per cent of the caracal's diet, while

Red hot pokers gleam amongst the relatively well-watered habitat in the higher ground of the Mountain Zebra National Park.

Karoo winters can be bitterly cold and the Mountain Zebra National Park is high enough for the occasional snowfall to bemuse grazing Cape mountain zebra.

RIGHT: *Cape mountain zebra sure-footedly make their way down a rocky slope.*

much of the balance is made up by rodents at 24 per cent and small livestock at 23 per cent. Therein lies the problem, with the accusation made that the Mountain Zebra National Park is a breeding ground and sanctuary for caracal that maraud the surrounding farming areas.

Moolman's work conclusively repudiated this perception. He estimated the total caracal population in the park at about 26, made up of perhaps four adult males, 10 adult females and 12 young. Surplus caracals are forced to leave the area and breeding data indicate a maximum of 12 young caracals should leave the park annually. However, in the district surrounding the Mountain Zebra Park, around 185 caracals were killed annually between 1975 and 1984. These figures indicate that the park's contribution to the caracal problem around Cradock is a minor one and also underscores the ability of this predator to survive despite every man's hand being turned against it.

As part of its policy of being a good neighbour, the Parks Board has done what it can to contain its caracal population; the entire boundary fence has now been electrified in an attempt to prevent caracal and other predators such as black-backed jackal from emigrating.

The Mountain Zebra Park is another National Parks Board ecological success story, having met its original objective of saving the Cape mountain zebra. With this battle won it seems certain that management will succeed in the drive to make this scenically attractive park become financially self-sufficient.

ADDO ELEPHANT NATIONAL PARK

That a naturally occurring elephant population could still survive just 70 kilometres from the major eastern Cape city of Port Elizabeth, and right in the middle of an intensively developed agricultural area, might seem something of a miracle. It could perhaps be so described, but the miracle would not have been possible without the Addo Bush – an unforgiving and virtually impenetrable tract of thorn-infested shrubland, unique to the lower Sundays River Valley and its environs.

To quote Dr Hans Grobler and Dr Anthony Hall-Martin of the Parks Board, the Addo Bush is an 'extensive thicket of evergreen, semi-succulent shrubs and small trees'. Renowned soldier and professional big-game hunter Major Philip Jacobus Pretorius, who was commissioned by the Government in 1919 to exterminate the elephants which were wreaking havoc on the crops, fences and water troughs of local farmers, had this to say: 'On the way down I visited the Addo, and soon realized that if there was a hunter's hell here it was – a hundred square miles or so of all you would think bad in Central Africa, lifted up as by some Titan and plonked down in the Cape Province.... It was scrub, generally some 18 feet high and exceedingly thick. Once in the jungle it was seldom possible to see more than five paces ahead and the jumble of undergrowth consisted of thorns and spikes of every description. A terrible country.'

Pretorius, of course, was prejudiced. He was trying to shoot the elephants and a number of unseen elephant at five paces stand a very good chance of killing the hunter instead. That prospect is enough to concentrate any hunter's mind closely on the problem at hand and one of Pretorius's more unusual innovations was to set up a ladder once he thought he was close enough to the hidden elephant and to shoot from the top of that.

LEFT: *An Addo bull deep in the Valley Bushveld which has protected and fed the elephant since the last century.*

Dracophilis dealbatis *belongs to the family* Mesembryanthemaceae. Mesembryanthemum *means midday flower and refers to its habit of opening in full midday sun and closing on dull, cloudy days and at night.*

Today's visitor will find the scene little changed from Pretorius's day and the dense vegetation along most of the access roads rarely permits one to see more than about five metres into the bush. For those who wish to get to grips with the thorns and spikes, it is possible to take a walk in the 400-hectare botanical reserve section of the park from which the elephant, buffalo and rhinoceros are excluded.

The best way to see the Addo elephant is to wait for them at one of the park's waterholes where the lengthy social rituals of drinking and bathing allow visitors a long and uninterrupted look at a herd. Even if one is fortunate enough to find a group of elephants in the road on a drive, the game-viewing experience here usually consists of watching their dust-covered rumps beating a steady retreat into the impenetrable bush for the aforementioned five metres after which they disappear.

Addo, in fact, has until recently been a highly frustrating park for the visitor because its other major attractions, black rhinoceros and buffalo, are also difficult to see. The buffalo are largely nocturnal in their habits while the rhinoceros remain in the thick bush throughout the daylight hours. However, as a consequence of the sweeping changes now being made in the Parks Board's administration, the park authorities have broken new ground with the introduction of night drives.

Such trips have been offered for years by several private game reserves because for many people they have proved to be the highlight of their bush experience. The newly instituted night drives at Addo make a world of difference because visitors not only improve their chances of seeing the elusive buffalo and rhinoceros – and of course elephant – but will also encounter a number of other shy or nocturnal creatures such as porcupine, bushpig, aardvark, bushbuck and kudu, as well as the smaller carnivores such as black-backed jackal, mongooses and genets.

The buffalo that now find refuge in the park are the last survivors of the herds that roamed the Cape Province before they were shot out by hunters and farmers in the 19th century.

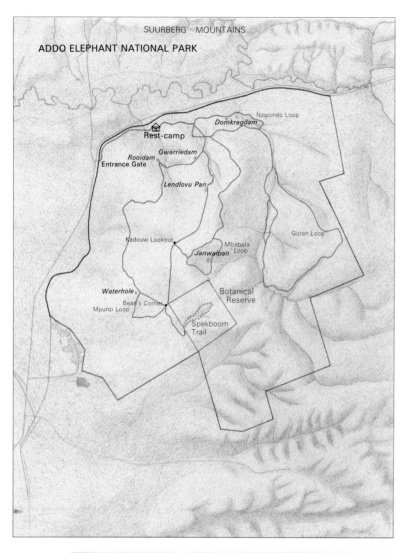

The flightless dung beetle feeds on the coarse dung produced by elephant, rhinoceros and buffalo. The ball of dung also provides a safe cocoon in which it lays its eggs.

Hunting pressure drove the survivors into the Addo Bush where their habits changed dramatically. Buffalo are normally predominantly grazers but their Addo Bush habitat often has very little grass. The buffalo have consequently been forced to become occasional browsers. They also tend to be less active by day than other buffalo populations elsewhere, perhaps to avoid exposure to hunters.

Grobler and Hall-Martin point out that the living conditions of buffalo in the Addo National Park are harsh compared with those in other parts of Africa. Typically, most Addo buffalo die before the age of 13 whereas buffalo elsewhere may live for 20 years. The population has been kept at around 60 animals since the early 1980s by the park management to cope with the effects of the severe droughts which this region experiences. A major effort has also been made to establish a buffalo population at Vaalbos National Park which now numbers over 50 animals.

The Addo buffalo herd was, and is, extremely important for the Parks Board and conservation in general because it is the only population in southern Africa free of the highly infectious foot-and-mouth disease so feared by cattle- and sheep-farmers. This means that surplus animals can be translocated from Addo to other parks and game reserves without upsetting the farming community and the veterinary authorities. In recent years buffalo from Addo have been transferred to the Willem Pretorius Game Reserve in the Orange Free State, the Andries Vosloo Kudu Reserve north of Grahamstown, the Vaalbos National Park and other reserves in Natal, Bophuthatswana, Swaziland and Namibia. Many others were sold to private game farms country-wide. Conditions for the buffalo at Addo have been greatly improved with the recent extensions to the park because about 800 hectares of the new ground is grassland.

Although black (or hook-lipped) rhinoceros were relatively common in the bush country of the eastern Cape when the white settlers first arrived on the scene, they soon fell victim to the hunter's gun. In fact the last known surviving specimen of black rhinoceros in the Cape Province was shot in this area in 1853. It was killed on the Coega River only 30 to 40 kilometres southwest of the Addo Park on what was later to become Sir Percy FitzPatrick's farm 'Amanzi'.

The black rhinoceros presently in the park, therefore, are not of original eastern Cape stock. Not only are they not local, they are the descendants of seven animals introduced from Kenya in 1961 and 1962. This population has thrived particularly well in Addo and now numbers more than 20. Their future in the park, however, is presently the subject of debate because they belong to the northernmost subspecies of the black rhinoceros in Africa (*Diceros bicornis michaeli*) and Addo was formerly the home of the southernmost subspecies (*Diceros bicornis bicornis*). At the time of their introduction, it was thought that subspecies *bicornis* was extinct, and the geographically closest subspecies, *minor* from Natal's Umfolozi and Hluhluwe reserves were not available for translocation to Addo: hence the purchase of rhino from Kenya. Thirty years later the situation has been turned around: Kenya's rhinoceros population has been reduced to about 400 animals, South Africa's 'extinct' subspecies has been 'rediscovered' in Namibia and, with rhinoceros poaching presently rampant throughout East and Central Africa, South Africa's well-protected conservation areas now hold out the best hope of survival for the black rhinoceros as a species.

The Kenyan rhinoceros in Addo are now of considerable importance to the conservation cause. Addo is excellent rhinoceros habitat and would make a superb park for the now rediscovered but still endangered South African subspecies *bicornis*. The National Parks Board now faces a dilemma: if it restores Addo to subspecies *bicornis* what does it do with the equally endangered subspecies *michaeli*?

The reason the area is so good for black rhinoceros and elephant lies in the quality of the vegetation. Although the veld is officially classified as 'Valley Bushveld', 90 per cent of the park is covered by a sub-type known as 'spekboomveld' or 'elephant's food' (*Portulacaria afra*), a bushy succulent-leaved shrub up to three metres in height. It is highly palatable with a high moisture content and it grows back quickly if not over-browsed. Although not apparent to the casual visitor, Addo plays host to one of the largest concentrations of big game in Africa in relation to its size. This is made possible by the nutritional richness of the spekboomveld.

Comparisons of veld productivity can be made by calculating the 'biomass' (the mass of animal and plant life a particular unit area can support). In 1978 the biomass of large grazing animals in the park was calculated at 4 800 kilograms per square kilometre which was substantially higher than that of any other national park or game reserve in South Africa. The biomass of the area was over 6 700 kilograms per square kilometre – the fourth highest in Africa.

One of the ironies of Addo's history is that although the spekboomveld was responsible for saving the elephant when they took refuge in it at the beginning of the century, the elephant in its turn may end up saving the spekboomveld. Over the years this veld type has been largely eradicated in the districts adjacent to the Addo Park to make way for agricultural expansion and the park now plays a key role in conserving the little that remains. Nearly 600 species of plants have been identified in the Addo Park and a small botanical reserve has been created within the park from which the elephants, buffalo and rhinoceros have been excluded to protect the different vegetation sub-types which are grouped under the term 'Valley Bushveld'. These are moist spekboomveld, bontveld, coastal bush,

Like Addo's rhino and buffalo, eland are seldom seen by day in the thick bush which provides virtually impenetrable shelter for all these large species.

and dry spekboomveld. The reserve also acts as a 'benchmark' for scientific study of the effect of browsing pressure by the large herbivore species on the rest of the park.

The story of the Addo Park is one of the great sagas of South African conservation history. By the end of the 19th century the only surviving elephants in the Cape Province were a handful of refugees in the fastnesses of the Knysna Forest and a somewhat larger number – perhaps as many as 150 – in the Addo Bush. The tract of bush 'ceded' to the elephants at Addo was left undeveloped by farmers because it was waterless and therefore valueless. Elephants, however, require over 150 litres of drinking water daily and the Addo herd was forced to move out of the bush every night to quench its thirst. During these excursions the elephants raided crops, killed livestock, damaged fences and water-points and were consequently hunted by the local farmers. The confrontation between elephant and farmer

Due to the highly palatable 'spekboomveld', the Addo Elephant National Park has one of the largest concentrations of big game in Africa in relation to its size.
LEFT: *The lengthy social ritual of drinking and bathing also helps the elephant to lose excess heat which, because of their great size, may often be difficult.*

resulted in Major Pretorius being commissioned in 1919 to exterminate the entire population.

He very nearly succeeded because he shot about 120 over the following 11 months. However, his all-out battle with the elephants attracted public attention and protests. By that time only 16 elephants were left. Over the next few years, despite the proclamation of Addo as a Provincial Elephant Reserve in 1926, no concrete steps were taken to safeguard the animals who continued to be harassed by long-suffering farmers, and when the Addo Elephant National Park was declared on 3 July 1931 there were just 11 elephants left. These hardy survivors had found refuge on the lands of Jack and Natt Harvey, the only farmers in the area prepared to allow the elephants to live unmolested on their property.

The Harvey property lay 30 kilometres from the newly declared park and the elephant had to be herded to their sanctuary by the first warden, Harold Trollope – a monumental task which took him three months. Water was provided from new boreholes in the park but for the next 20 years a running battle between the elephants and the local farmers continued because the elephants were still able to leave the park at will to raid nearby crops. The farmers retaliated by shooting them whenever they caught them on their lands and by 1954 the total Addo population had increased only slightly and stood at just 18.

The solution was finally found by warden Graham Armstrong who, between 1951 and 1954, perfected his elephant-proof fence consisting of steel bars made from old tram lines and steel wires in the form of old elevator cables. He proved its success when, in a trial of strength, elephants were unable to break through it to get at oranges placed within a protected enclosure. The cables were donated by the Waygood Otis lift company, the posts were cut from old tram-lines dug up in Johannesburg and Port Elizabeth, while the deep holes for the posts were drilled by a machine on loan from the Post Office. Extra finance was provided by The Wildlife Protection Society and the Port Elizabeth Publicity Association. The fence was officially named the Armstrong Fence in honour of its inventor.

Initially an area of 2 270 hectares was fenced in and the elephant population has risen constantly since then to the current level of around 200. The fenced in area now totals 11 726 hectares and a similar barrier has been erected around the new ground at a cost of R81 000 a kilometre in 1992 money values.

As might be expected, given the way in which they had been harried all their lives, the original Addo survivors were exceptionally irascible elephants with an understandably hostile attitude towards humans. The authorities therefore deemed it necessary to restrict access by tourists to the park for many years. That visitors can today drive safely throughout the park is thanks mainly to Dr Anthony Hall-Martin – now Executive Director: Southern Parks.

In 1976 he undertook a research project on the Addo elephants but found the extremely aggressive nature of the animals a severe handicap to his work. However, he faced down a number of charges from aggressive and mistrustful cows and bulls, gradually forcing them to accept, firstly, his vehicle and then other vehicles, so that eventually tourists could drive safely in the elephant enclosures.

One of the problems facing the Parks Board at Addo is the genetic danger of inbreeding in the elephant population. The entire herd is descended from the founder stock of 11 in 1931 and is therefore seriously inbred to start with. The scientists rule of thumb to reduce the deleterious effects of inbreeding is that the population should be over 500 strong. At its present size Addo is simply not large enough to allow a herd as large as this.

Addo has a population of the valuable michaeli *subspecies of black rhino. These were originally imported from Kenya and have thrived in the park.*

Over the years, therefore, the Parks Board has attempted to acquire extra land, with some success. From 7 735 hectares in 1981 when it supported 108 elephants, the park increased its size to 9 175 hectares in 1990 with 135 elephants. A further recent purchase extended the area to 12 126 hectares. The latest additions of land were made possible by money raised through the Rhino and Elephant Foundation (of which Dr Anthony Hall-Martin was a co-founder with Clive Walker and Peter Hitchins) and the Southern African Nature Foundation.

Ideally the Board would like to purchase much more ground, depending on the availability of funds, to increase the park by as much as another 50 000 hectares. The long-term hope is to extend Addo eastwards towards the village of Paterson and north to link up with the Zuurberg National Park.

A distinctive feature of the Addo elephants is that the cows are virtually tuskless while the bulls have small tusks compared with those of the Kruger National Park population. This is usually attributed to the intensive hunting pressure in the 19th century which was directed against the large-tusked animals, resulting in the genetic characters for large tusks being shot out of the population. However, the park authorities point out that most of the elephants in Addo are still rather young, the majority of them having been born since 1954. As the lifespan of an elephant is no more than 60 years, it is possible that a number of the bulls may yet develop large tusks as they grow older.

Visitors to the park are handed a notice at the entrance gate which asks them not to run over any dung beetles in the roads. The flightless dung beetles of Addo may be small in size, but they are just as fascinating to the discerning observer as are the elephants. Once widespread throughout the Western and southern Cape, this beetle which specializes in feeding on the coarse dung produced by elephant, rhinoceros and buffalo, has suffered a severe decline with the disappearance of the pachyderms which provided its food source. Thanks to its elephants, Addo is home to the largest known surviving population of this rare and interesting insect.

The request to drive with care was necessitated by the elephant's predilection for walking on the roads at night. There they deposit their dung and there the beetles tend to concentrate. As a result of research done on the beetles, warden Lucius Moolman says Addo's roads will be graded in a different fashion to leave a gentler slope on the shoulders. The work showed that beetles rolling their balls of dung were unable to negotiate the existing steep shoulders on the roads and were trapped along them in large numbers, often dying of heat exhaustion.

The rest-camp at Addo was greatly expanded at the end of 1990 from just six rondavels to 24 chalets and Moolman says management attention is focused on making the park financially viable by attracting visitors year-round. While it is full over

Lichen, commonly known as 'old man's beard', is found draped on trees in the forested sections of Addo.

holiday periods, the new camp had an average occupancy rate of only 50 per cent for 1991. The Board plans to build conference facilities which will attract conventions to the park and Moolman wants to start advertising Addo nationwide in conjunction with the other eastern Cape parks. He is also looking at such innovations as a children's playground as this 'gets around the problem of bored children which means the adults will be able to spend longer in the park. It fits in with the new policy of the Parks Board which is to budget money for eco-tourism projects that will make money.'

In the past Addo has been derided by critics who charged that it was too small to be ecologically viable, that it could never be a serious tourist attraction and – worst of all – it was a despised 'single species' reserve set up to conserve an insignificant population of an otherwise abundant species. These criticisms have now been overtaken by events. With its recent land acquisitions, and proposals for further expansion, Addo Elephant National Park is gradually becoming a part of South Africa's valuable network of ecosystem reserves.

ZUURBERG NATIONAL PARK

This park, at present virtually unknown because the general public has so far had very limited access to it, should become one of South Africa's major wildlife attractions when the Parks Board completes its development programme and opens up the area to tourists.

Although only about 35 000 hectares in extent at present, Zuurberg National Park is destined to grow considerably larger in the near future in order to make it more viable in ecological terms. It is a splendidly scenic landscape of rugged, jumbled mountains and steep river valleys situated about 50 kilometres north of Addo Elephant National Park and extending westwards along the Suurberg Mountain Range towards the Sundays River.

The park's array of plant and animal species is impressive, a consequence of its richly varied habitats. Nevertheless its large mammal fauna has suffered over the last 200 years and the Parks Board will have to reintroduce certain species including elephant, buffalo and black rhinoceros.

The birdlife of the area is particularly diverse. Amongst many others there are the lovely Narina trogon and the crowned eagle in the Valley Bushveld and forest patches, jackal buzzards and Cape rock thrushes on the mountain slopes, and orange-breasted and malachite sunbirds along with Cape sugarbirds in the fynbos. The magnificent black and martial eagles are a common sight as they soar effortlessly over the park.

Much of the Zuurberg has been protected since 1896 when it was proclaimed a conservation area by the Department of Forestry. On its acquisition by the National Parks Board in 1985 it covered some 20 000 hectares in three separate sections; these have subsequently been linked through land purchases.

Although not yet proclaimed, some of the land between the present western boundary of the park and the Sundays River has recently been acquired by the National Parks Board from

A female Cape rock thrush with her young.
These birds are common in mountainous areas.

the Department of Water Affairs. This has given the park control over more than six kilometres of both banks of the river and allowed the introduction of a small herd of hippopotamus towards the end of 1992. The Board hopes to acquire further land along the river in the near future.

The Board's intention is to zone the Zuurberg for various recreational uses. The highest ground will be declared a wilderness area through which a five-day hiking trail will run; tents will be erected by the Parks Board for the four overnight stops. There will be a separate zone for day-hikers as well as for horse-trailing. Another area, perhaps as large as 12 000 hectares and incorporating the Valley Bushveld, forests and some of the grasslands, will be fenced off as a big-game area for the eventual reintroduction of elephant, buffalo and black rhinoceros and tourists will be able to go into this area on guided trails.

At present tourists can only pass into the Zuurberg Park at two points and stay overnight at just one of them. A guesthouse which can accommodate six people has been built at Cabouga on the western side of the park; this is reached by a gravel road from the town of Kirkwood. The house is set in a magnificent natural amphitheatre created by the surrounding mountains and visitors can explore these on either a short or a long trail. Bookings are made through the warden at Addo Elephant National Park and not through the usual central booking offices in Pretoria or Cape Town.

On the eastern side of the park, which is reached by taking the gravel road signposted 'Zuurberg' from the entrance to the Addo Elephant National Park, several day-trails are available to visitors; all of these trails start from the warden's office.

While the biological diversity of the Zuurberg Park and its ruggedness are two of its greatest attractions, they also pose major management problems for the ecologist. The Suurberg Mountain Range is part of the Cape Fold Mountain Belt and the geological contortions of its Witteberg quartzite rock strata are one of the dominant scenic features. The topography is extremely rugged with deep valleys although there are no dis-

LEFT: *The rugged slopes and high country of the Zuurberg National Park provide breathtaking scenery for hikers.*

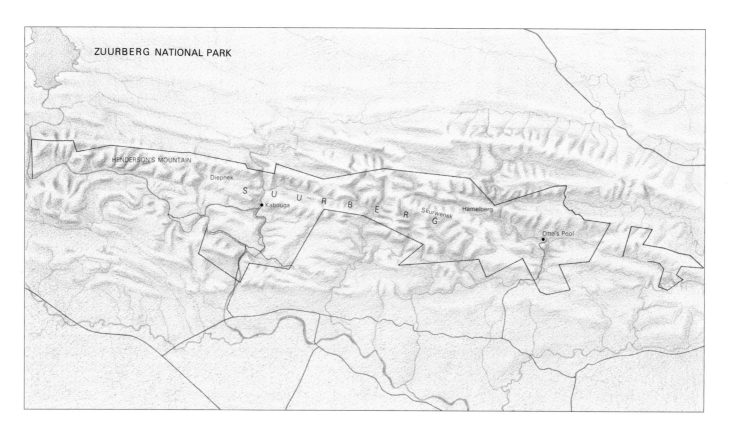

tinct peaks or very steep cliffs and the height above sea level varies between only 250 and 970 metres.

There are five major vegetation types in the park. Grassy fynbos covers about 33 per cent of the area with Valley Bushveld accounting for 32 per cent, grassland 18 per cent, Afromontane forest 12 per cent and mountain fynbos five per cent. These different vegetation types are found close together in a patch-work distribution. It is a delight for hikers because the rugged nature of the terrain means frequent changes of scenery as the trail winds upwards out of the Valley Bushveld and forests of the valleys to the grassland and fynbos of the mountains. The proportion of fynbos to grassland varies according to whether the mountain slopes are north- or south-facing with the true fynbos usually on the steeper, colder and wetter southern slopes.

Scattered throughout the mountainous parts of the park are the magnificent cycads which occur either singly, jutting out of a vast expanse of grassland, or in colonies, standing proud on the ridges and steeper slopes. Some superb specimens of Thunberg's cycad have attained heights of over four metres. The three cycad species found in the park are regarded as threatened species: the Karoo cycad (*Encephalartos lehmannii*) is 'critically rare', while Thunberg's cycad (*E. longifolius*) and the eastern Cape dwarf cycad (*E. caffer*) are both 'vulnerable'. Other threatened or rare plant species include *Euryops latifolius*, a chincherinchee *Ornithogalum anquinum*, the Suurberg cushion-bush *Oldenburgia arbuscula*, and the Cape krantz ash *Atalaya capensis*.

These plants are numbered amongst the most valuable assets of this national park, but their conservation, along with the many other species in the various plant communities, provides the park warden with vexing problems. For one of the main management tools used by ecologists on three of the veld-types in the park is fire in the form of controlled burning, and the different veld-types demand very different burning régimes.

Considerable research has been carried out on the problem and the conclusions are that grassland should be burnt every three years, while the minimum period between fires in grassy fynbos should be about five years and in mountain fynbos about 12 years. Just how to achieve this ideal is the predicament facing the Zuurberg managers as it will be difficult, if not impossible, to restrict controlled burns to specific types of vegetation given the patchy nature of its distribution.

The Afromontane forest patches present a further problem. Although they themselves are resistant to fire, their margins are not, and successive fires could gradually erode the forest patches through destruction of trees on their perimeters.

According to Parks Board research officer Dr Peter Novellie, 'a burning policy can therefore not be formulated purely on the basis of published information derived from apparently similar

RIGHT: *In many parts of South Africa, aloes such as this* Aloe ferox *brighten up winter vegetation with their brilliant coloration.*

situations in other localities. Long-term burning trials on the Zuurberg will be needed to determine the optimum programme for our particular situation.'

In the meantime, it has been decided that the conservation of mountain fynbos will take priority over the maintenance of grassy fynbos which in turn has priority over grassland; the justification for this policy is that agricultural land in the southern and eastern Cape fynbos areas is generally managed to encourage grassland at the expense of fynbos by burning frequently. The Parks Board intends to reverse this process and err on the side of fynbos by burning less frequently.

Novellie points out that fires occur naturally in the Zuurberg. They are caused by lightning and there is evidence that they tend to occur at five-year intervals on average. This, he feels, may have been the major factor influencing the distribution of the fynbos communities in the park. Fire every five years is too short a burning cycle for mountain fynbos but the fact that this veld-type exists on the wetter and steeper southern slopes of the Suurberg suggests that it survives there because this part of the park is less prone to fire and may not burn every five years.

Another management problem is the question of how the introduction of large herbivores like buffalo and elephant will affect the area. The size of the populations that can be permitted will be determined by the primary conservation objectives of the Zuurberg Park which are, firstly, to maintain the diversity of plant communities, and secondly, the diversity of plant species within each of these communities.

Cape mountain zebra have already been reintroduced to Zuurberg from the Mountain Zebra National Park and Karoo National Park and at the end of 1992 there were around 20 in the herd. The Parks Board's intention is to bring back red hartebeest and eland as well. Novellie recommends the introduction of small populations of these animals at first so that their movements and patterns of habitat use can be carefully monitored; ecologists will be able to detect changes in the vegetation in relation to animal density and thus determine maximum population limits for each species to prevent any deterioration in veld quality or loss of species diversity.

Once these ecological problems have been satisfactorily overcome there is no doubt that the Zuurberg National Park will be one of the top environmental attractions in the country, for where else in South Africa, or the world, could you go hiking through fynbos and game-viewing for four of the big five – elephant, black rhino, buffalo and leopard?

RIGHT: *A number of different vegetation types are found in the Zuurberg Park, ranging from fynbos on the mountainous tops to Valley Bushveld and Afromontane forest in the ravines and gulleys.*

TSITSIKAMMA NATIONAL PARK

Living inland, flying routinely from continent to continent in a matter of hours, it is difficult to appreciate the power of the sea and the enormous control it exercised on the world's lines of communication before the development of steam-driven ships and then aircraft.

The sea has been tamed to an extent by technology but its raw power is evident to any visitor to the Tsitsikamma National Park who happens to be there when a big swell is running. Waves up to three and four metres high, generated by storms thousands of kilometres offshore in the Atlantic and Indian oceans, roll inexorably landwards to smash against the cliffs that plunge abruptly into the sea along this stretch of coast.

This is a rocky shoreline and there are no waves lapping gently on golden beaches to lull the senses. Instead there is a constant assault from the deafening roar of the waves tearing themselves apart on the outlying reefs and sending huge spurts of shattered surf and foam soaring into the sky. A wild sea is a very real experience at Tsitsikamma's Storms River Mouth Restcamp where the log cabins and the 'oceanette' flats stand close to the edge of the cliff where salt water rages at jagged rock. In fact, four of the first chalets built at Tsitsikamma were destroyed in 1978 by waves from a freak storm. Reconstruction took place a prudent distance further inland.

The feeling of being totally exposed to the elements, while marked at Storms River, can become overwhelming on the park's famed Otter Trail during a storm where the hiker makes his way between the raging sea on the one hand and the rocky hillsides rising sharply inland on the other.

Despite these harsh conditions the area teems with life although this is perhaps not so readily apparent, being hidden below sea level for much of the time or frustratingly camouflaged in the forests that cover the ridges inland. The Narina

The Knysna lourie may be seen in the dense thickets of the Tsitsikamma Forest. Its rich crimson wing feathers are displayed only in flight.

LEFT: *Fynbos in full flower on top of the cliffs that plunge into the Indian Ocean on the Tsitsikamma coast.*

trogon hoots hidden in the tree canopy while the rustle of dead leaves ahead of the hiker could have been a blue duiker, scuttling away without showing itself.

The landlubber can potter around the edges of the sea peering at the life in rockpools but those able to snorkel or use scuba equipment can get a far better look at the ocean life along the park's underwater trails. Still, exploring around the edges involves getting a closer look at one of the most fascinating sections of the park, the intertidal zone. This is the region of reefs and rock-pools which is flooded twice a day for about six hours by the tides and then left exposed to the vagaries of the elements for the other 12 hours.

Fossicking around in rock-pools is one of the traditional pursuits for families who holiday by the sea but the vast majority probably have no conception of the astonishing adaptations that have been forced on the animals and plants found in these pools to cope with this robust environment.

It is, as Margo and George Branch point out in their book *The Living Shores of Southern Africa*, one of the most stressful of habitats for animals and plants. While submerged, the zone is subject to cold water and pounding wave action and, when exposed during the day, to temperatures of up to 40 degrees Celsius; as a result the uncovered animals can lose as much as 70 per cent of their body water. In order to withstand these circumstances 'all the vital functions of life such as respiration, excretion and reproduction must be adapted to function in two completely different environments: marine at high-tide and essentially terrestrial at low-tide'.

The different levels of rocky shores can be divided into zones depending on the extent to which they are flooded and exposed during the tidal cycles. On South Africa's south coast the furthest up the shore is called the Littorina zone and is so named for the tiny *Littorina africana* snails which cling to the rocks in their thousands. This region is submerged only during high tides and supports a very limited variety of life because of the scarcity of food brought in by the waves. Food becomes more

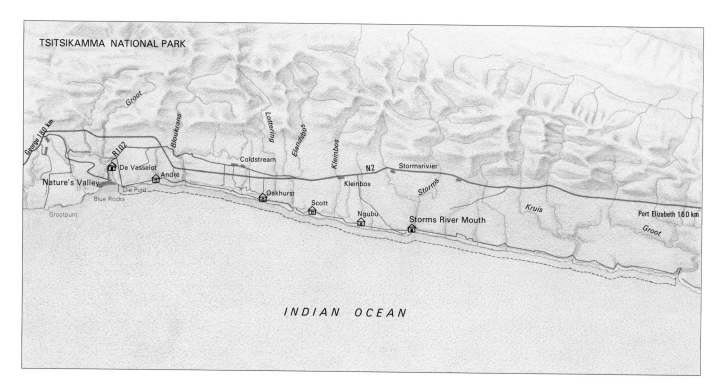

abundant further down the shore and the size of many of the animals increases accordingly.

The next zone seawards is called the Balanoid zone and is so named after the scientific name of the barnacles which characterize it, *Balanus*. This area is inundated twice a day by the normal tides and it is split into two sections, the Upper and Lower Balanoid zones. The difference between these zones and the Littorina zone is immediately to be seen in the numbers and variety of molluscs in the former, such as whelks, winkles and limpets. Along this stretch of coast there are nine species of limpet including the giant *Patella tabularis* and the beautifully marked *Patella miniata*.

Limpets minimize water loss and survive the period when they are exposed to the air by sealing their shells against the rock face through adhesion. At night, limpets move slowly over the rocks grazing on algae and then return each day to the same spot on the rock face. Their shells eventually develop until they fit precisely into the outline of this home site which helps to preserve the limpet's near-airtight seal.

The Lower Balanoid zone is exposed only at low tide and is marked by thick beds of algae. Sea anemones are common here and prey on a variety of small creatures, including fish, by catching them with their waving tentacles, each armed with stinging cells to immobilize their victims.

The next zone down is the Cochlear zone, named after the limpet *Patella cochlear*, which forms a dense band at the low-tide mark on most south coast rocky shores. These limpets may reach a density of up to 2 600 per square metre, the highest density

achieved by any limpet of this size elsewhere in the world. However, along the rugged Tsitsikamma coast the Cochlear zone has been modified by the pounding wave action, and brown mussels tend to dominate this zone with *Patella cochlear* found in large numbers only in the more sheltered spots. Where these limpets do occur they are often so densely packed that the juveniles are forced to live on the backs of the adults, probably the only place where the larvae can settle and not be eaten. The brown mussels anchor themselves to the rocks with tough horny byssus threads and grow in such dense colonies that nothing else can live amongst them.

Below the Cochlear zone and thus below the low-tide mark lies the infratidal zone. The most commonly seen citizen of the infratidal zone is the so-called red-bait, a species of sea squirt which grows in dense colonies. The orange-red flesh from which it gets its name is hidden by a black, leathery outer covering and these animals are greatly sought after by fishermen who use them as bait. Also common and readily seen in the infratidal zone are various types of starfish, as well as several species of algae and, in sheltered areas, large numbers of sea urchins which feed on the algae.

The animals of the intertidal zones have been shaped through evolution to fight moisture loss and high temperatures but some have evolved better designs than others. A limpet has an inherently bad design for heat gain and loss because the flat shell exposes a large surface area to the sun while its large 'foot', which the animal uses to cling to rocks, also allows a rapid gain of heat from sun-heated rocks.

Winkles are better shaped to cope with heat as they have taller, more rounded shells and a smaller 'foot'. This is why *Littorina africana* can survive so high up on exposed shores.

The animals in this fascinating world are divided up largely into grazers, filter-feeders and predators. Winkles and limpets are grazers living on seaweed. Mussels, red-bait and barnacles are filter-feeders – they extend frond-like arms which trap tiny particles of food that are swept over them by wave action and so filter their food from the ocean. Such filter-feeders dominate on rocky shores where they can readily find anchorage.

Predators come in three forms: they may sit and wait for their prey like anemones; actively hunt them down as in the case of starfish; or they may scavenge on dead and weakened prey as whelks do. In their micro-world whelks are a force to be reckoned with. Species like *Nucella dubia* are able to drill a hole through the shell of a limpet and then feed on the soft animal within using a long proboscis. One remarkable species of whelk, *Argobuccinum pustulosum*, actually produces concentrated sulphuric acid which it pumps into the nest-holes of its favourite prey, the reef-worm *Gunnarea*; it waits a while for the worm to dissolve and then sucks up its meal.

A starfish preys on mussels by hunching over them and relentlessly applying pressure from its multitude of feet to pull open the mussel's valves. As the mussel resists this attack by clamping the two halves of the shell tightly together, the starfish then turns its stomach inside out over the mussel and pours digestive juices containing a muscle relaxant over it; this causes the valves to open and the digestion of the mussel to proceed.

Among the more interesting predators in this area are two species of spiders which have adapted to life in the intertidal zone. These go hunting at night during low tides then, when the tide comes in, return to a cell or nest which they have constructed in the empty shell of a limpet or barnacle; the nest traps an air bubble which provides the spiders' oxygen supply until the next low tide. Spiders caught by the incoming tide before they can return to their nests can breathe under water because their waxy body hairs trap a thin film of air around them which acts as a reservoir of oxygen; furthermore, as the spider uses up the oxygen, more diffuses from the water to the air film across the surface of the bubble.

Interesting though the adaptations of such small creatures may be, however, they are not the main attraction for the average visitor to the coast who is usually more interested in fishing. Tsitsikamma is a marine-fish reserve but anglers are allowed to fish along a three-kilometre stretch of the park's coast; their catch records form a key part of the research programme under way in the national park.

According to research officer Nic Hanekom, work in the Tsitsikamma Park has been aimed at determining the ecological importance of the park as a marine reserve which is surrounded

A rock crab feeds on minute particles of food with delicate movements of its huge pincers.

Sea urchins flourish in the rock pools along the coastline and are easily found at low tide.

Scarlet coral formations are one of the many wonders awaiting scuba divers and snorkellers on the Tsitsikamma underwater trails.

by areas in which fish are heavily exploited by both recreational and commercial anglers. The initial work was carried out on three fish species which are important for both sport and commercial line-fishermen – the roman, dageraad and red steenbras. It revealed, as might be expected, that the population densities and sizes of these fish are greater inside the park, but also that the reserve benefited the surrounding areas by 'seeding' them with fish that had grown up in the reserve. Further research is in progress to quantify this seeding effect.

Another important conservation feature of the park which has only recently come to light is that it is a breeding reserve for the shallow-water squid known as 'chokka' which has become an important resource for commercial line-fishermen in recent years. The main breeding grounds for these squid have been shown to lie in the 200-kilometre-long stretch of coastal waters between Plettenberg Bay and Port Elizabeth; as the Tsitsikamma National Park conserves some 70 kilometres of this coastline, the protection of such a large part of the breeding ground must be crucial to the economic wellbeing of the chokka industry. The chokka's name undergoes a transformation by the time it reaches the consumer: it is sold under the name 'calamari'.

Perhaps the most graphic illustration of the wealth of life at sea off this coastline occurs on the rare occasions when mass movements of pilchards or anchovies take place and fortunate

LEFT: *Low tide exposes stretches of mussel- and limpet-encrusted rocks along the shoreline.*

is the visitor who is in Tsitsikamma when this happens. Warden Dennis Bower describes it as 'a moving food chain'.

The huge shoals of small fish come close inshore followed by schools of dolphin and predatory fish such as tuna and yellowtail which feed on the shoals. The dolphins and tuna in turn are harried by sharks while at the same time the pilchards are being constantly dive-bombed by squadrons of Cape gannets out to share in the feast.

So far, 13 species of dolphin and whale have been recorded in the marine reserve with very large schools of common dolphin and bottlenosed dolphin often being seen. Other species present include striped and humpback dolphins. Whales are regularly sighted from the shore although a combination of patience and keen eyesight is required. One of the most important species is the southern right whale which moves from sub-Antarctic waters to the southern coast of Africa during the winter months to breed. Southern right whales have mated and calved in the mouth of the Storms River. Adults reach lengths of about 18 metres and can weigh up to 67 tons.

With its long stretch of Cape coastline, Tsitsikamma National Park is extremely vulnerable to man's misuse of the marine environment – in particular to the pollution of the oceans by plastics and oil dumped or spilled by vessels at sea. Mammoth tankers each carrying between 250 000 and 400 000 tons of crude oil pass round the southern tip of Africa daily as they ferry the oil from the Persian Gulf to consumers in Europe and North America.

The potential for a disaster of enormous proportions is constantly present if a major oil slick from one of these vessels should drift on shore. A review of South African maritime history shows there have been a number of close escapes. The most recent was in 1991 when the fully laden tanker *Atlas Pride* had its bows smashed in by huge waves in South African waters and was forced into Algoa Bay where, after a few nerve-racking days, its cargo was safely transferred to another tanker.

Tsitsikamma Park's closest call came in June 1979 when the freighter *Evdokia* ran aground on a reef about 100 metres off the national park's shore and spilled about 440 tons of bunker oil. Fortunately the oil slick was carried offshore by the currents and little pollution of the coastline took place.

Bower says there is absolutely nothing that could be done to prevent the widespread ecological devastation to the intertidal zone that would occur in the event of a huge oil slick washing ashore at Tsitsikamma. All he and his staff could do is trust in the dynamics of the rocky shore to break up and wash away the pollution as quickly as possible.

Plastic pollution does not threaten the same scale of ecological disaster that can be caused by a major oil spill but it is an insidious and persistent menace in the ocean, killing large numbers of sea birds and marine mammals. It also litters the most remote and pristine beaches where capricious tides and currents bring the debris ashore. At Tsitsikamma, the area most affected is the aptly named Driftwood Bay near the Storms River mouth.

According to the Dolphin Action and Protection Group of Fish Hoek near Cape Town, surface trawls in South Africa's coastal waters have shown densities of about 3 500 particles of plastic per square kilometre. Although this is considerably less than densities recorded for the Northern Hemisphere, there is no room for complacency. The danger is that the trend is upwards and the nature of the pollution, for example, boat oil-containers and discarded fishing nets, points directly at the culprits who are recreational and commercial fishermen and boat-owners. Nylon ropes, nets and line are particularly dangerous because the material does not rot and break down in the water like ordinary rope. Marine mammals such as dolphins and fur seals have been killed by becoming entangled in nets and line while many species of sea bird eat small plastic particles which will eventually kill them.

The sea is not all that the park has to offer, however, as the forested areas are an attraction in their own right. These have been greatly enlarged with the recent acquisition of the 2 560-hectare De Vasselot section. This area encircles the little holiday community of Nature's Valley at the western end of the

LEFT: *The Cathedral Rock arch at the mouth of the Keurbooms River is a landmark on the Tsitsikamma coast.*

Red leaves are highlighted against the green background of an overgrown stream flowing through the forest.

park and the acquisition benefits both the village and the park. The pristine forested areas around the village are now under the best environmental protection South Africa can offer and, as Nature's Valley cannot now be extended, the park's western flank is secure from the inappropriate property development which has spoiled so much of the rest of the Garden Route.

The National Parks Board acquired the then De Vasselot Nature Reserve from the Department of Forestry in 1987 through an exchange for the 478 hectares of forest that previously formed the Tsitsikamma Forest National Park. This latter ground included a rest-camp near the Storms River Bridge which the Forestry Department has since sold.

In 1992 a further 20 000 hectares of fynbos-covered mountains was added to the park through a 30-year contractual arrangement with Rand Mines Properties (R.M.P.). Negotiations are also under way to add a further 16 000 hectares of ground to Tsitsikamma by taking in the stretches of State Forest situated between De Vasselot and the R.M.P. ground, with a view to joining them into a consolidated national park stretching some 20 kilometres inland from the coast.

The De Vasselot section is named after a Frenchman, Comte M. de Vasselot de Regné, who in 1880 was appointed Superin-

tendent of Woods and Forests for the whole of the then Cape Colony. Under him the first real efforts towards conserving the Tsitsikamma forests were made in an attempt to reverse the excessive exploitation that had taken place in the area since 1836.

Tourist facilities at De Vasselot are adequate but basic, and Bower says they will be kept that way because that's how visitors prefer it. The camping-site on the banks of the Groot River has 45 plots but no electricity. The particular attractions of the De Vasselot section are the six different day trails, ranging in duration from two to six hours, which wind through the forests and along the rivers and beaches of the area.

The dominant features of the forest walks are the towering giant yellowwoods, two species of which occur in the park. The larger of the two species is the Outeniqua yellowwood which can reach a height of 60 metres in the Knysna Forest but not as high at De Vasselot. It is distinguished from the other species, the real yellowwood, by its characteristically flaky bark and its leaves which are thinner and shorter than the leaves of the smooth-barked real yellowwood.

These two species are, however, only part of the plant wealth of these forests which contain 122 woody tree and shrub species as well as a host of lichens, fungi, ferns, mosses and orchids.

A waterfall is formed by a small river plunging over a ridge as it drains the well-watered mountain ranges lying inland from the coast.

Other important tree species – which are identified by labels on many of the hikes in the national park – include stinkwood, candlewood, Cape chestnut, assegai, ironwood and Cape beech.

Lichens and fungi are prominent and colourful members of the forest community with the most common lichen being *Usnea barbata*. This unusual and interesting plant is better known as 'old man's beard' and is commonly seen hanging in long tatters from the crowns and branches of large trees. Lichens are strange composite plants consisting of two organisms, an alga and a fungus. The alga obtains water and essential elements from the

fungus which in turn lives on the food that the alga manufactures through photosynthesis.

Most fungi are plants, but some seem to combine the characteristics of both plants and animals. Like animals, fungi obtain their energy from organic compounds which they get from living or dead plants or animals. They are like plants in that they cannot move but differ importantly in that they are incapable of photosynthesis – the essential mechanism by which the vast majority of plants create organic compounds from the sun's energy, water and carbon dioxide. Fungal fruiting bodies can be spectacular, some 'bracket' fungi forming saucer-like structures coloured anything from off-white to glowing orange.

However, the most noticeable aspect of the Tsitsikamma forests for the visitor is the dearth of large animals. Research has attributed this to the nutrient-poor soils in the southern Cape forests with all the available nutrients having been extracted by the plants and kept for their sole use in what is termed a 'closed nutrient cycle'. The trees have evolved mechanisms to deter animals from feeding on them, one such being the production of chemical compounds in the leaves to make them unpalatable. The closed nutrient cycle is extremely efficient, and as soon as a fallen leaf rots in the leaf litter on the forest floor its nutrients are snapped up by a network of fine surface roots sent upwards into the litter by the nearest tree.

One of the mammals that does occur in the Tsitsikamma forests is the diminutive blue duiker. Adult males of this species stand a mere 30 centimetres at the shoulder and weigh just four kilograms. These interesting little duikers are currently being researched by Tsitsikamma's Nic Hanekom in conjunction with Zimbabwean zoologist Viv Wilson.

Many duiker populations of a number of species are under pressure throughout Africa but a handful of these essentially forest-dwelling antelope, including the blue duiker and Maxwell's duiker of West Africa, have shown a surprising ability to live in close proximity to man. They can survive in secondary habitats close to human activity and, on this protein-starved continent, that means they have the potential to be an important source of meat if managed properly. In West Africa, duiker have always been a major source of food, being sold as 'bushmeat' at a higher price than beef, mutton or goat. However, a number of more sensitive duiker species have been brought to the verge of extinction in the process.

The blue duiker has a very scattered distribution, stretching from the Cape northwards through the forest zones of Africa to Nigeria, but it has not been studied to any significant extent. The work in Tsitsikamma has shown that blue duiker population densities are low in the southern Cape forests compared with the Natal coastal forests. The reason appears to lie in the poorer nutritional quality of the food and in a less abundant fruit supply in the Cape forests.

Leopard and caracal occur sparsely in Tsitsikamma but the park's best-known predator, and one which visitors have a slightly better chance of seeing, is the Cape clawless otter after which the Otter Trail – of which more later – is named.

Two species of otter occur in South Africa. The spotted-necked otter is completely restricted to fresh water and seldom moves out of a riverine or lacustrine environment. The Cape clawless otter, however, is far more adaptable and will go seafaring in estuaries and along rocky coastlines such as at

The Cape clawless otter is usually seen in the early morning or late evening. It forages for prey, mainly crabs and octopuses, in rock pools and river mouths.

Tsitsikamma and often wanders quite far from aquatic environments in the interior of the country. However, access to fresh water is essential to its survival.

It can reach a length of 1,5 metres and a mass of up to 19 kilograms. The forefeet have no claws but two or three digits on the hindfeet have rudimentary nails. Although the toes of the hindfeet are webbed for half their length, the forefeet are adapted for feeling and grasping and are only slightly webbed. Despite the lack of webbing in comparison with the spotted-necked otter, the Cape clawless otter is an excellent swimmer using its long, thick tail to assist in propulsion.

Researchers at Tsitsikamma have estimated an otter population density of one for every two kilometres of coast, giving a total of 35 to 40 otters for the park east of Nature's Valley. Approximately 120 holts, or resting sites, of otters were found along the coastline, always located in dense bush and usually associated with a supply of fresh water.

Radio-tracking has shown the otters to be mainly nocturnal with peak activity occurring between 8 p.m. and 10 p.m. when the animals forage mainly along the submerged rocky ridges close to shore. The population density of otters in Tsitsikamma is high and is attributed to the plentiful supplies of food and

Bracket fungi spread their saucer-like rings on a piece of dead wood in the Tsitsikamma Forest.
LEFT: *The canopy of the tall forests creates a cool, quiet haven from the sun and the powerful sea breezes.*

fresh water as well as the safe haven provided by the existence of the national park. Examination of otter 'spraints' or droppings has shown 81 per cent of the prey taken here consists of four species – two types of crab, octopus and the suckerfish. The rest of the prey is made up of more than 30 species of which at least 20 are fish.

The park's famous Otter Trail has come to be regarded as South Africa's premier hiking trail. Its popularity is such, however, that prospective hikers have to book a year in advance to obtain a place. It takes five days with four overnight stops to cover the 41 kilometres, and is not an easy trail as it involves frequent lung-sapping climbs and knee-jolting descents along the rugged coastline. Just to make it really interesting there are two river crossings where hikers will have to swim if they arrive there at high tide.

If it all sounds far too much like hard work for a holiday, Tsitsikamma offers a series of easier day trails around the Storms River mouth, as well as at De Vasselot. There is an excellent boardwalk which now allows elderly visitors to undertake the one-kilometre walk from the Storms River camp restaurant to the Strandloper archaeological site. This is a cave where early hunter-gatherers sheltered while living in the area.

This superb national park, which was South Africa's first marine reserve, has something to offer every visitor which no doubt accounts for its popularity with both local and overseas tourists.

KNYSNA NATIONAL LAKE AREA

The sequence of events concerning the Knysna National Lake Area amounts to a serious setback for South African conservation and should have set the alarm bells ringing for the Knysna community as well as for other coastal towns facing similar circumstances.

After taking over the lake area from the Lake Areas Development Board in 1983, the National Parks Board announced during 1992 its intention to relinquish responsibility for the ecological management of the Knysna estuary.

There were a number of reasons – which will become abundantly clear in this chapter – but essentially the Parks Board felt that the lake area did not qualify as a national area in terms of recognized I.U.C.N. criteria as it is only a small part of a much larger system. An additional reason was that in practice it could not satisfactorily carry out its conservation mission given the forces ranged against it that were intent on developing the area.

A revealing pointer to the problems facing the Knysna National Lake Area is the fact that warden Peet Joubert maintains that his most valuable ability is being able to read a set of building plans. He acquired the knowledge from one of his various occupations before turning to nature conservation and, ironically, this is what he now spends a substantial portion of his working hours doing instead of the tasks normally associated with nature conservation. He has also acquired an in-depth knowledge of the various ways in which domestic sewage can be disposed of, for this is a matter of crucial importance to the environmental health of the Knysna Lagoon as the population around its fringes continues to expand.

Knysna, one of the most scenic parts of the Garden Route and ecologically the richest estuary in the southern Cape Province, also contains some of the highest-priced properties in the country. Indeed, a property boom of unprecedented propor-

A Knysna lily glows fire-red in the forests surrounding the lagoon.

tions is in progress as developers respond to the ever-increasing demands of wealthy holidaymakers and retired people – and make a substantial profit for themselves in the process. Apart from the extensive development planned around the lagoon itself, another proposal is to build a new town the size of Knysna just east of the lagoon on the coast at Noetzie.

Such developments place Knysna in the front line of the war between environmentalists and the more insensitive developers and Peet Joubert now finds himself caught in the crossfire: 'I balance enforcing the law to the letter against the resentment this will cause among the residents; this is not a "normal" national park but what one could call a "people-park" – and that involves negotiation and compromise. However, I come down immediately and as hard as is required on anything that seriously threatens any of the lagoon's vital ecosystems. I am constantly in trouble with environmentalists because I don't stop development, and with the developers because I'm always interfering with their plans. I am not against development but developers must accept ecological considerations in their projects. Frankly, I think they would get more money for projects that were sited correctly and surrounded by natural Knysna forest or fynbos than they do from their normal approach which is to clear everything away and create artificial surroundings.'

Joubert is not despondent about the situation, for he believes acceptable solutions can be found. Often they are quite simple, for example persuading every householder to install a 1 000-litre tank to catch rainwater would go a long way towards solving the looming freshwater crisis. The lagoon has a limited catchment area and the growing population is putting greater demands on it. 'I'm an idealist but if I can get a realist to meet me halfway then we can get something done. But I suppose I'm also like a mother-in-law – I never stop moaning', he remarks.

From a conservation point of view the situation is certainly far from ideal and the task facing the National Parks Board is a particularly difficult one because this is a 'recreational' park. While the Board has the power to enforce certain regulations, it

LEFT: *The Knysna Lagoon fills up at high tide as the Indian Ocean surges in through the Heads.*

KNYSNA NATIONAL LAKE AREA

The shoveller crayfish is identified by its pink and white stripes and distinctive 'shovels'.

does not have the final say on many other aspects on which it can only comment or advise. Building regulations are laid down in municipal bylaws and can be changed by application to the municipal authorities. A number of developments that have included features opposed by the Parks Board have in fact been allowed to proceed at Knysna.

As in the Wilderness National Park property-developers propel their projects to the limit of the law and frequently overstep it. Joubert points out that a number of property erven scheduled for development are, in fact, submerged by lagoon waters for six months of the year. Problems have arisen from the fact that the Parks Board claims jurisdiction over the lagoon waters up to the high-water mark together with the land under the water. Unfortunately, there is a certain amount of overlap between the Sea-shore Act (No. 21 of 1935) and the Lake Areas Development Act (No. 39 of 1975), both of which bear upon the development of ground and the 'reclamation' of land under the high-water mark. As certain landowners possess tracts of land around the lagoon which were marked out and purchased before these acts were promulgated, they feel they have the right to 'reclaim' land at the water's edge. Joubert disagrees vehemently, and believes that the owners have no right to fill in areas below the high-water mark without the approval of the Parks Board.

Such filling in of the margins of the lagoon, which can be seen at a number of recent developments, destroys sections of the most important component of the lagoon ecosystem, the salt marshes. These function both as the 'battery' for the system, generating food and providing shelter for its teeming life, and as an essential filter by trapping pollutants such as silt and sewage entering the lagoon. Joubert estimates that perhaps 20 per cent of the lagoon's salt marshes and other associated wetlands have been lost during the past 30 years through property development and the building of causeways for roads and railway-lines across parts of the lagoon.

Salt marshes, like other types of wetland, can cope with and purify certain levels of polluted waters and in fact this capacity has been put to good use with the development of artificial reed-bed sewage filters. Over a critical level, however, their purifying capacity is overwhelmed or they are buried under silt. Joubert describes silt in rivers as the 'ecological equivalent of throwing sand in your gearbox'.

Apart from its size – the lagoon is some 1 800 hectares in extent – the important feature of Knysna is that its lagoon is permanently open to the sea because of the deep mouth between the two famous 'Heads'. Strong currents pushing between the two rocky headlands ensure the mouth does not get closed by longshore drift as is the case at Wilderness.

For this reason Knysna became one of the first ports in the country, commencing in 1776, when the Dutch East India Com-

A young rockcod peers inquisitively at a diver. Lagoons provide essential shelter for these fish when immature.

As one of the top South African game fishes, this young leervis will be eagerly sought after by anglers once it has matured.

pany set up a timber station to exploit the Knysna forests. The port closed down in 1953 when its operations became unviable. Only ski-boats and yachts now accept the challenge of navigating the channel between the Heads, often to be greeted once in the open sea by dolphins playing in the heavy swells. Those swells combined with hidden reefs and powerful currents can make the passage between the Heads a tricky exercise.

According to Margaret Parkes and V.M. Williams, co-authors of *Knysna – The Forgotten Port*, 44 ships have met their end along this stretch of coast. Many came to grief trying to get into or out of the lagoon and the remains of one – the 460 ton sailing ship *Paquita* which sank in 1903 – lie close to the parking area adjacent to the restaurant at the Heads.

The fish life of the lagoon is particularly rich and diverse because of the permanent link to the sea which keeps the salinity of water in the lagoon close to that of sea water at 35 parts per thousand. The salinity of the lakes in the Wilderness system gets diluted by river flooding when the estuary mouths are closed and can drop to as low as six parts per thousand.

Some 200 species of fish have been recorded in the Knysna Lagoon of which about 81 species are wholly or partly dependent on this and other South African estuaries for their survival in South African waters. The Knysna sea horse is one of eight species of fish totally dependent on estuaries for their existence and the Knysna Lagoon hosts the largest remaining population of this sea horse which has an extremely limited range. It is

found only here, in the Keurbooms estuary and at Swartvlei and Mossel Bay. A much commoner, larger species of sea horse occurs from East London northwards up the coast from which the Knysna sea horse population has developed in isolation.

Fish found in the estuary include many of the marine species most sought after by anglers and this popular sport is one reason for the large and growing numbers of boats on the lagoon. Control of the numbers of boats and a zoning of the lagoon for recreational activities are being negotiated by the National Parks Board with affected parties in an overall management masterplan for the lagoon due to be finalized in 1993.

Some 22 species of fish including top sporting species like white steenbras, spotted grunter, leervis and Cape stumpnose could not survive in South African waters without access to estuaries where the juveniles grow to maturity. Marine populations of a further 28 species including anglers' favourites like kob, elf, blacktail and giant kingfish are boosted because these fish are able to utilize both the open sea and the estuaries.

The preferred way of angling for many of these fish is to use bait in the form of mudprawns, sand-prawns and bloodworms which form part of the mass of life that teems in healthy salt marshes and mudflats. All these animals are known as 'detritus feeders' – they live on the rich food soup created by bacteria breaking down the plant detritus in the water.

According to Margo and George Branch's book *The Living Shores of Southern Africa*, each sand-prawn may produce three

The grey heron is commonly seen wading or standing motionless for long periods in shallow water as it waits for potential prey to come within striking distance.

An African spoonbill collecting nesting material. These birds forage by sweeping their partly submerged bill from side to side while wading in shallow water.

grams of faeces a day, all of it rich in organic matter and thus also food for bacteria. In the case of one small estuary in the western Cape for which figures are available, sand-prawns have been shown to add more than 200 tons of faecal material to the food chain each year; the total for the Knysna Lagoon must exceed this figure substantially.

Ecologists have therefore felt it necessary to protect the prawns in the Knysna Lagoon from overexploitation. Out of season, there are already about a thousand people taking bait on a regular basis from the lagoon; in season this pressure increases enormously. Each angler is allowed to take a maximum of 50 prawns a day for bait but Joubert admits control of this is difficult. His main concern is not so much over the number of prawns taken as the damage done to the mudflats by the number of bait-collectors trampling over them. Digging for prawns is illegal because of the damage this does but even the legal method of getting them out of their holes with a prawn-pump harms the mudbanks.

One sector of the lagoon has been set aside as a 'bait reserve' where no collecting is allowed. Joubert would like to introduce to Knysna the system used in Natal and in areas such as Port Alfred and Swartkops near Port Elizabeth where the public are

The water of the lagoon reflects the bronzed sky as the sun sets over the Knysna National Lake Area.

not allowed to collect bait; instead official bait-collectors catch and sell bait organisms to fishermen as and when required.

Three key elements in the management master plan for the lagoon are, firstly, control over access to the lagoon through recognized launch-sites; secondly, control over the number of boats on the water; and thirdly, the generation of revenues for the Parks Board. At present the Board derives no income from the Knysna National Lake Area but is spending more than half a million rands annually in direct running costs to manage it.

It has been proposed that a boat-licensing system should be introduced, with a differential tariff for residents and non-residents, and also a tariff charge per kilowatt of engine power to penalize the large, fast boats; their more powerful engines can cause serious ecological damage from the wave action created by their wakes. The public is likely to kick against such restrictions but the harsh reality is that such measures will have to be taken. Otherwise, Knysna will not retain the natural attractions which motivated so many people to build retirement and holiday homes here. Bringing it off successfully requires the warden to add tact and the patience of Job to his talents as a plumber, building inspector, environmental policeman and ecological jack of all trades.

The Knysna sea horse is dependent on the Knysna estuary for its entire life cycle. It is therefore particularly susceptible to any damage to its environment.

WILDERNESS NATIONAL PARK

The Wilderness region at the western end of South Africa's world-famous Garden Route matches up to the vision that many people have of a paradise on earth. Lushly forested mountains sweep down into a coastal plain spangled with a chain of picture-book lakes whose waters flow into the azure Indian Ocean through a magnificent golden, sandy shoreline.

It is a paradise for the sun-lover, surfer, hiker, fisherman and bird-watcher, as well as for those who simply wish to relax and unwind by the sea. How to keep it that way is the challenge faced by the National Parks Board which took over management of much of the area in 1985. And unfortunately there is no shortage of problems in this particular paradise.

The threats to Wilderness come from several sources: from property developers (altruistically) desirous of meeting the soaring demand for housing and time-share schemes from people who wish to live here; from the huge and growing influx of visitors who swamp the area in peak holiday seasons; from the looming confrontation over water resources, which must meet domestic requirements as well as the needs of the entire lakes ecosystem; and from a Sword of Damocles hanging over the area in the form of a proposed freeway system planned to go straight through it.

The situation is complicated by the structure of the national park itself and also by the decision by the National Parks Board to 'soft-pedal' in its approach to enforcing park regulations, at least in the short term. The Parks Board has in fact bent over backwards in an effort to accommodate the wishes of the local residents who agreed in 1985 to allow the area to be proclaimed as a national park. Prior to this the Wilderness area was controlled by the Lake Areas Development Board which operated under different rules and considerably less stringent legislation.

The strikingly coloured purple gallinule may be found in dense reedbeds from which it occasionally emerges to feed in open marshes.

The problem appears to be that a number of residents, some of them extremely well connected politically, would like to have a national park on their doorstep to protect and preserve the Wilderness environment, but without the tiresome restrictions that go with it, such as control over boat-launching sites and restrictions on the exercising of dogs.

One specific case occurred late in 1991 when South Africa's Electricity Supply Commission (Eskom) constructed a power-line through a section of the national park adjoining Langvlei. There was an alternative route higher up in the adjacent hills but the farmer owning the ground refused to allow the pylons to be erected on his property. Strangely, the National Parks Board agreed to let Eskom use its ground.

The Wilderness National Park has a major problem with its boundaries caused by the piecemeal development of the area in the years preceding its proclamation. Stretches of national park lie cheek by jowl with private ground, public roads, a railway-line, State land earmarked for the proposed freeway, and areas controlled by the Cape Provincial Administration (Goukamma Nature Reserve) and the Southern Cape Regional Services Council.

Along the frontal dune system east of the town of Wilderness, residential plots set just back from the edge of the dune adjoin the national park. In 1991 one resident, eager to create a sea view from his newly built house, cleared the bush between it and the edge of the dune. In so doing he cleared a section of the national park and cut down a number of milkwood trees. Apart from the fact that *all* vegetation within a national park is sacrosanct under the National Parks Act, the milkwood is a specially protected tree under the Forest Act.

The Wilderness Park is also divided between two different magisterial districts, George and Knysna. 'Taken overall, we have too many boundaries and a number of them are not sufficiently well defined', comments warden Paul Sieben.

Legal actions brought by the National Parks Board were under way in late 1991 against certain property owners and developers to solve disputes over the boundaries at Swartvlei

LEFT: *An aerial overview of the Wilderness coast, lakes and Touw River.*

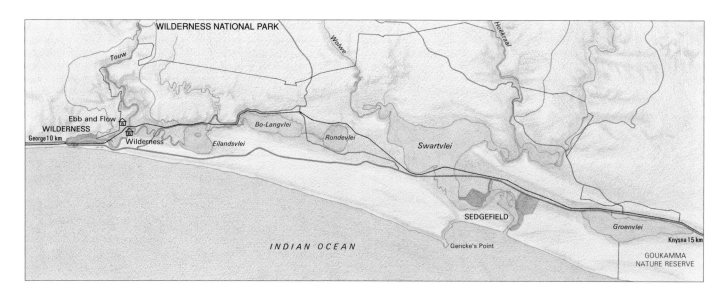

where the two magisterial districts of George and Knysna meet. Swartvlei lake itself is part of the Wilderness National Park but the land adjacent to it is privately owned, although it falls into the 10 000-hectare 'Control Area' inherited from the Lake Areas Development Board. Boundaries are defined as the 'edge of the lake' in one district and as the 'high-water mark' in the other. Both definitions are lacking in clarity and property-developers on the lake-shore are taking advantage of this to encroach on the

A hike in the mountains above Wilderness reveals flowers such as these ericas, which indicate that the region is part of the Cape Floral Kingdom.

lake itself. Infill is dumped along the water's edge so that they can build more chalets and other holiday-orientated structures as close to the water as possible.

Such encroachment has a number of adverse effects. Not only does it reduce the size of the lake but it damages or destroys the ecologically important reedbed and water-weed areas around the lake margins. Vegetation in this zone to a depth of three metres accounts for 86 per cent of the primary food production in Swartvlei, forming the base of the food pyramid for the invertebrates, fish and birds that live here. Most holiday-resort developers clear away sections of reed and weedbeds for boat-launching sites as well as for swimming areas.

The erection of so many buildings almost at water level has directly and detrimentally impinged upon the most crucial ecological management decision affecting this unique water ecosystem, that is, when to open the mouths of the two estuaries connecting the lakes to the open sea. Both the Swartvlei mouth and the Touw River mouth at Wilderness have to be opened artificially at regular intervals using heavy earth-moving equipment. This is necessary because the outflow is not strong enough to prevent them from being closed by 'longshore drift', whereby currents running parallel to the coast gradually pile sand across the mouths. Dropping sea levels over the past 7 000 years have tended to isolate the lakes from the sea to the point where only during major flood periods would the two estuaries be opened naturally to the ocean. The consequences of the flooding of the margins of the lakes before a natural opening of the mouths occurs are not acceptable because of the extent of building development around the shores of the lakes.

Artificial opening of the estuaries can only be satisfactorily achieved when the lake waters have risen to the point where a strong flow of water to sea will take place through the newly opened channels, thus scouring them out deeply. If the mouths

Low-lying clouds carrying the threat of imminent rain create a moody, grey landscape at Langvlei Lagoon.

are opened prematurely they will be closed prematurely as the longshore currents will push vast volumes of sand back into the estuaries and clog them up. However, there is a fine line between ensuring a good flow to sea by allowing the lake levels to build up, and at the same time avoiding flooding of lake-shore properties. If the building of more houses directly on the edge of filled-in stretches of lake-shore is allowed, it can only result in yet more pressure on the Wilderness park warden to open the mouths prematurely.

Development in the Wilderness area, as well as around Knysna and Plettenberg Bay, is controlled by a Guide Plan drawn up in 1983 by the then Department of Constitutional Development and Planning. These areas were assessed by planners and zoned for various purposes such as recreation, nature areas, industrial development, township development, agriculture and forestry. Changes to the zoning arrangements can be applied for and a number have in fact been granted since 1983. Some property developers, however, appear to be proceeding with their projects without waiting for the necessary approval: unfortunately they seem to be getting away scot-free.

The increase in the number of residents of the town of Sedgefield has also adversely affected environmental conservation throughout the entire Wilderness region. Sedgefield has grown greatly in recent years because of the demand for holiday-homes: this rapid expansion has been supported by the Municipality and the Regional Services Council, both of which gladly accept the extra income from rates and taxes.

The trouble of course is that each new development further erodes the natural attributes of the Wilderness region which the newcomers found so appealing in the first place. Foremost amongst these attributes are the lakes that make up the Wilderness National Park: Eilandvlei, Langvlei, Rondevlei and Swartvlei.

The first three form the Wilderness Lakes System which is connected to the Indian Ocean through the floodplains of the Touw and Duiwe rivers. The much larger Swartvlei is a separate system and has its own estuary which reaches the ocean near the town of Sedgefield. Swartvlei is bisected by a railway-line; its larger section north of the line is 8,8 square kilometres in extent, while the estuarine section south of the line covers 2,2 square kilometres.

Immediately inland from Wilderness the forested slopes of the Cape Fold Mountains rise towards the interior of the continent.

The fifth lake in the region is Groenvlei which adjoins but is not part of the national park; it is completely cut off from the sea and its waters are the least brackish of the five.

The lakes have been formed by the rising and falling levels of the sea over the last 20 000 years. As the sea level fell, ridges of sand dunes were formed along the coast and behind them the lake depressions were created and filled in either by rivers or by flooding when the sea rose again. The process is still under way with the lakes gradually becoming shallower and shallower while their links to the sea are becoming more difficult to maintain. If the present processes continue unchanged, the lakes will eventually be taken over by reedbeds and disappear. That, however, should take another few thousand years, and of course if the polar icecaps melt as a result of the prophesied 'greenhouse effect', the sea level will rise and the lakes will be given a new lease of life.

Preliminary studies by botanists from Rhodes University have recorded 320 different plant species in the vegetation in and surrounding the five Wilderness lakes. Another study showed that six of the most important waterplants in Eilandvlei, Langvlei and Rondevlei produced about 2 000 tons of organic

material annually, forming a substantial food base for the Wilderness Lakes System. This material includes both the live plants which can be eaten, and the material from dead plants which can also be eaten by detritus-feeders or which breaks down through decay and returns nutrients to the water.

The lakes are extremely important as nursery areas for a number of fish species. Some of these are essentially marine fish which are hatched at sea but which enter the lakes through the estuaries and spend anything from one to several years in their rich feeding-grounds. Some 56 species of marine fish have been recorded in Swartvlei and 26 species in the Touw River system. The spring and early summer months appear to be the best times to open the estuary mouths to allow the young fish to enter the lakes and the older fish to leave. Oddly enough, most estuarine fish species do not spawn in estuaries but migrate out to sea for this purpose. Very soon after hatching, however, the fry return to the abundant food and protection which is offered by the estuarine environment.

More than 270 species of bird have been listed for the region which is arguably the key waterbird sanctuary along the southern Cape coast. Bird counts have shown more than 2 000 duck of nine different species on Langvlei at one time, and also a total of 7 000 birds representing a truly remarkable 65 species of waterbird at another. The park is home to such rare species as the osprey and a notable feature of Langvlei is the flocks of up to 100 great crested grebes that gather here at times.

The core of the Wilderness National Park is formed by the Rondevlei and Langvlei lakes which are totally under the control of the National Parks Board with clearly demarcated and fenced-off boundaries. They are the only parts of the park to which access can be controlled and are its most important bird and fish sanctuaries.

The Parks Board recognizes that the wealth of birds on these two lakes is one of their major attractions and intends capitalizing on this. Rondevlei, which has a superb bird hide, became part of the national park in 1987 when the Cape Provincial Administration agreed to relinquish control of its Lakes Nature Reserve. The Parks Board has built another hide on Langvlei and plans to construct several more.

Hiking is the other pursuit which the Parks Board would like to encourage and, so far, four trails have been opened up in the immediate vicinity of the camps at Wilderness with a fifth at Rondevlei. These trails range in length from three to 11 kilometres. The Pied Kingfisher Trail on the Touw River has a beautifully constructed boardwalk which allows hikers to walk through the reedbeds along the river without damaging the environment. Negotiations are in progress with local property owners to make at least one of the trails into an overnight excursion. All are day trails at present.

The priority, however, is to achieve the targets which were laid down in the 1983 Guide Plan. This had the overall objective of ensuring that further physical development in the area would take place without harming its rich variety of ecologically sensitive features.

However, there is much to be done at Wilderness before this aim becomes reality. The structure of the park and the aims of the Parks Board need to be examined carefully because the present situation in which most visitors flout the rules with impunity is a dangerous one if allowed to continue.

Parts of this region are a national park in name only and perhaps the worst example is the recreation area along the Touw River just off the national road at Wilderness. This is little more than a boisterous public picnic-site where people walk dogs, litter, make a great deal of noise and, in general, do what they please. The nearby beach areas are not much better.

Short of a draconian crackdown which the Parks Board clearly wants to avoid, it seems unlikely that the general public will ever treat these areas with the respect they deserve as integral parts of a national park. It can be argued that there is very little to conserve there anyway, but the danger is that the people who

treat this section as a general recreational area, may try to do the same thing in the more ecologically sensitive areas such as the beaches near Gericke's Point.

The National Parks Board will have to lay down the law, and do so clearly, as to what it will allow in which sectors of the Wilderness National Park. If it fails to do this, it may find that it is fighting a losing battle to conserve the region's natural beauty. It has pursued a deliberately lenient course since taking over responsibility for the area in 1985, in an attempt to gain the confidence and respect of the public while at the same time trying to educate people in the basic principles of nature conservation. Nevertheless the inevitable confrontation has merely been postponed. Chief Executive Director Dr Robbie Robinson counters by pointing out that the Wilderness region meets I.U.C.N. criteria for national park status and, despite the many problems, the present situation is, as he says, 'better than nothing. We have a say over development plans and can appeal to the authorities to stop certain developments if we feel this is necessary. It's an uphill and very frustrating exercise but still worthwhile if we can achieve a compromise between development and environmental conservation. It takes time to persuade people to alter traditional patterns of behaviour, but I really do believe that eventually we will get the public to accept and be enthusiastic about this very beautiful national park.'

A southern right whale breaches off the coast at Wilderness. Whales are seen regularly during the summer months.

BONTEBOK NATIONAL PARK

Precisely when can a formerly endangered species be considered safe? The question could be of crucial importance for the future of the tiny Bontebok National Park near Swellendam in the south-western Cape. Proclaimed in 1931 to prevent the last few remaining bontebok from following the footsteps of the blue antelope into extinction, the number of bontebok has risen from the original seriously endangered 17 to about 200 which is the maximum number the 2 786 hectare park can support without inflicting serious damage to its various plant communities.

Slow but sure, an angulate tortoise moves through his habitat in search of food.

Although the park itself can only support a limited number of bontebok, over the years its surplus stock has been disposed of to other nature reserves and private landowners with suitable habitat and the total world population now stands somewhere between 2 000 and 3 000. The exact total is not known because of the transfers of bontebok out of the south-western Cape region and the issue has been further complicated by cross-breeding between bontebok and the closely related blesbok. The bontebok has been taken off the *South African Red Data Book – Terrestrial Mammals* list of endangered species but is still classified as rare.

This has led to considerable speculation over the future of the Bontebok National Park with some conservationists voicing the opinion that the park has fulfilled the aims for which it was set up and should now be deproclaimed and run as, perhaps, a game reserve under the Cape Nature Conservation (C.N.C.). It is felt by some that the National Parks Board should turn its attention to areas more deserving of the highest degree of environmental protection that can be offered. Specifically, it has been suggested that the nearby De Hoop Nature Reserve, currently run by the C.P.A., should be upgraded to national park status. De Hoop, together with the adjacent Overberg test site, now has between 400 and 450 bontebok and is the jewel in the Cape

Provincial Administration's conservation crown. Given the rivalry between the Parks Board and the C.P.A. Nature Conservation Department, it is understandable that the C.P.A. intends resisting the loss of De Hoop.

A counter-argument forcibly put by Bontebok National Park warden, Otto von Kaschke, is that the bontebok cannot be considered safe because of concern over genetic impurity as a result of cross-breeding with the blesbok, and the very limited number of large – above 100 – herds of bontebok. The Cape Point Nature Reserve, run by the Western Cape Regional Services Council, which had a large number of bontebok, has reduced the numbers by selling them off, an action which Von Kaschke finds hard to understand. He claims that 'the bontebok's situation can by no means be described as well-off. Any population below 1 000 individuals is in trouble and a major die-off in population bringing numbers down to those levels could happen at any time through disease.'

Concern over the genetic purity of the bontebok has arisen because of ill-advised hybridization ventures by game farmers, particularly in the northern Cape and Orange Free State, of the bontebok with the blesbok. The two are closely related sub-species but the populations became separated during their development with the bontebok becoming a fynbos endemic restricted to a very small range in the south-west Cape, while the blesbok remained on the plains of the northern Cape, Orange Free State and the southern Transvaal.

Since 1988, the Cape Provincial Administration has started a register of proven 'genetically pure' bontebok and imposed restrictions on translocation of the animals. The National Parks Board now sells its surplus stock from the Bontebok Park only to approved buyers.

But the Bontebok National Park itself is more than just a haven for this handsome subspecies of antelope: it is also an island of coastal renosterveld, perhaps the most threatened vegetation sub-type of the Cape's world-famous biome. Intensive agricultural practices in the form of wheat-'and sheep farms

LEFT: *Bontebok graze on the plains below the towering Langeberg Mountain Range near Swellendam.*

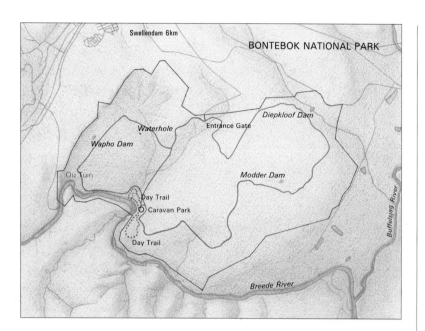

number of mammals including grey rhebok, red hartebeest and a group of recently reintroduced Cape mountain zebra. The area around Swellendam once hosted all the 'big five' – elephant, rhinoceros, buffalo, lion and leopard – but only the leopard now remains, and only in the mountains. The traveller is reminded of their occurrence by road signboards such as 'Buffelsjag Rivier' (buffalo hunt river). 'Renosterveld' was so named because of the numbers of rhinoceros encountered in this particular veld type by the pioneers.

The bontebok was 'discovered' by European settlers as they trekked eastwards from Cape Town in the late 17th century. It was not an abundant antelope even then, however, and from travellers' diaries and other written accounts it appears to have been restricted to the coastal area bounded by the Langeberg range to the north, the Hottentots Holland Mountains near Caledon in the west, and the Riversdale district in the east. The restricted range of the bontebok was what nearly brought about its demise once the first European hunters and settlers moved into this area as it inevitably fell easy prey to the hunter's gun, and by the middle of the 19th century there was a real possibility that they could become extinct. There was indeed an uncomfortable precedent: the blue antelope, whose range coincided more or less with that of the bontebok, was completely exterminated by the year 1800 – the first historically recorded African mammal to become extinct.

have almost totally eradicated the indigenous vegetation from the Swellendam/Bredasdorp region, leaving the park as one of the few refuges for seven rare plant species, including three species of *Aspalathus* and one *Diosma*. Von Kaschke feels that there is a pressing need for thorough plant surveys to be done in the park and one of his aims is to set up a 'fynbos trail', giving visitors the opportunity to enjoy a long hike through the park in addition to the short two-kilometre trails which already exist.

The island of conservation formed by the Bontebok National Park is home to 11 species of frog, 17 species of reptile and a

The bontebok owes its survival to the foresight of a number of families in the Bredasdorp area who, from about 1864, tried to protect the animals on their lands. By 1927, however, bontebok numbers had dropped to an estimated 121 animals and the Union Government was forced to intervene. It purchased the farm Quarrie Bos near Bredasdorp in 1930 specifically to conserve the bontebok and in 1931 this farm was proclaimed as the Bontebok National Park. The choice of ground was not a good one, however, because much of it was prone to flooding and, although the bontebok numbers increased, the animals suffered severely from parasitic infections caused by the damp conditions. Many of the animals were also noted to suffer from 'swayback', a weakness in the hindquarters caused by a lack of trace elements, particularly copper, in the grazing.

The National Parks Board tried unsuccessfully to find more land adjacent to Quarrie Bos but were able to purchase the present Bontebok National Park on higher ground near Swellendam. The 84 bontebok then on Quarrie Bos were transferred to Swellendam in 1960 and the new park was proclaimed in 1961. The translocated animals thrived in their new home, reaching a maximum population of 320 in 1981.

A Cape sugarbird searches for nectar on a king protea. Its diet consists mainly of insects which are often caught in the air.

RIGHT: *The Breede River forms the boundary between the Bontebok National Park and the wheat farms which surround it.*

Bontebok are closely related to blesbok and interbreeding between the two by some game farmers is a threat to the survival of the bontebok.

Situated in the Cape Floral Kingdom, the Bontebok National Park is home to numerous spectacular flowering plants such as this species of Leucadendron.

Buffalo were originally stocked in the park but were removed because it proved impossible to keep them in and trespassing animals wreaked havoc in the surrounding wheatfields. Von Kaschke would like to see both bushbuck and eland reintroduced to the park. Bushbuck occur on nearby farms and he feels that the park is large enough to support a small eland herd.

Seldom seen but still present are the clawless otter, water mongoose, aardwolf and caracal while some 200 species of birds have been recorded. These include South Africa's national bird, the blue crane, and the impressive Stanley's bustard which presents such a spectacular sight in the summer months when the males display their magnificent orange and white plumage to potential mates.

At present plans are underway to expand the park and to make it more attractive to visitors to this part of the south-western Cape. Many tourists overlook the Swellendam area which is halfway between the two main landmarks on their maps – Cape Town and the Garden Route.

The National Parks Board is therefore to acquire an additional 500 hectares of ground from the Swellendam Munici-

Grey rhebok bound across an open stretch of fynbos towards cover.

pality. This land, formerly used as the municipality's camping-ground, will link the existing park to the N2 national road just south of Swellendam. Here the Board plans to build chalets, a restaurant and a conference centre to supplement the small camping facility presently available.

This development will help to alleviate the chronic shortage of tourist accommodation around Swellendam and the ready access off the main road will make the facilities more attractive to passing tourists. The Parks Board also plans a promotion campaign for the entire Overberg region, stretching from Hermanus in the west to Swellendam in the east, to be run on a joint basis with other organizations such as the various municipalities and the Cape Nature Conservation. Such a campaign would help the Bontebok National Park to become financially viable and independent which it is not at present.

When all these projected plans come to fruition, the Bontebok National Park, which began as a lifeline for one endangered subspecies of mammal, will have expanded its role to take its rightful place as an integral and important part of the south-western Cape's tourism network.

South Africa's national bird, the blue crane, is commonly seen both in the Bontebok National Park and on the surrounding farmlands.

RESEARCH, DEVELOPMENT
AND ADMINISTRATION

Research, according to Chief Executive Director Dr Robbie Robinson, is the foundation upon which the work of the National Parks Board rests. For without the knowledge provided by detailed ecological and biological studies, it would be impossible to manage the many different ecosystems protected by South Africa's network of national parks.

The Board will therefore continue to spend heavily on research despite the constraints being imposed on it in the effort to become financially self-sufficient.

Dr Robinson insists that the Parks Board's research effort must be management-orientated with specific goals for the programmes agreed upon in advance between the researchers and the wildlife managers.

He does not believe that the Board 'can afford the luxury of doing research simply for its own sake. That will have to be done by institutions other than the National Parks Board to whom we will continue to grant access to the various parks to allow them to carry out their research.'

Nevertheless, financial pressure is being brought to bear on the Board's own research budgets and many essential projects being undertaken by its staff could not be carried out without funding assistance from non-government conservation organizations such as the Endangered Wildlife Trust, the Southern African Nature Foundation, the Wildlife Society of Southern Africa and the Rhino and Elephant Foundation – to name but a few. Help is usually in the form of cash grants to pay for necessities such as petrol or the cost of helicopter flights, as well as in the provision of equipment such as vehicles, radio-collars and tracking apparatus.

The Endangered Wildlife Trust, for example, has given substantial sponsorship through the Stuart Bromfield Fund to the

In the Addo Elephant National Park the rare, flightless dung beetles have right of way.

Kruger Park's specialist scientist Dr Gus Mills for his project on African wild dogs. Mills acknowledges that 'were it not for the Stuart Bromfield Wild Dog Fund we would be in serious trouble trying to complete this project'. The Trust's particular interest is the wild dog, the most endangered large carnivore in Africa.

The Kruger National Park's research manager Dr Willem Gertenbach believes the research programme in any national park has six main objectives.

The first is to compile an inventory of all the biotic components of the park – mammals, birds, reptiles, amphibians, fish, invertebrates and plants – before proposals can be drawn up for their proper conservation. Considerable research effort goes into these inventories which can be very time-consuming, particularly when dealing with the myriad plant and insect species as well as marine life.

The second stage is the detailed study of each ecosystem to determine how the various forms of plant and animal interact with one another, while the third stage involves 'troubleshooting' – identifying existing or potential ecological problems which the park managers may have to deal with.

The fourth stage requires the research team to make management recommendations based on its findings and the fifth objective is to ensure that they are properly implemented. Implementation of specific measures also includes a monitoring function by the researchers to see whether they are having the desired effects on the ecosystem.

The sixth and final stage is to ensure that the knowledge gleaned from the programme is made available to other interested parties in order to prevent unnecessary and wasteful duplication of effort when others are faced with similar problems. Most of the Board's research results are published in its own scientific journal *Koedoe*.

An example of a well-structured research programme is the Kruger Park's river project which has developed into one of the largest multidisciplinary research programmes in South Africa

LEFT: *A helicopter guides a herd of springbok towards the waiting nets in a game-capture operation.*

Handling a young elephant darted for translocation is a labour-intensive operation.

with 42 registered projects being carried out by 80 researchers. The findings of this programme will be applicable to any other river system in Africa, and probably the world, and the National Parks Board will be responsible for their dissemination.

Over the years the Parks Board has developed a detailed system for the successful management of its wildlife resources, particularly in arid savanna regions like the Kruger National Park and the Kalahari National Park. These range from the determination of appropriate veld-burning regimes for grass- and bushveld to the pioneering of sophisticated game-capture systems involving the development of gas-powered dart-guns and immobilizing drugs. Monitoring takes up a large part of the scientific programme. This entails detailed aerial surveys during which game numbers are counted and vegetation conditions are recorded.

The Board's scientists have also introduced innovative veterinary techniques to safeguard rare animal species against disease. One example is the vaccination of roan antelope – classed as 'endangered' in South Africa – against anthrax by darting

them from helicopters with a 'bio-bullet' of vaccine that is harmlessly absorbed into the animals' bodies.

As of the end of 1992 the National Parks Board's research team consisted of 28 researchers, five technicians and 33 supporting staff. In addition there were 130 'outside' researchers working in the various national parks where 245 registered projects were underway.

Many research projects produce results which are at variance with conventional and long-held views about certain animals. A prime example is the work by various researchers on the spotted hyena which has shown it to be a formidable predator, second only to the lion. This is in sharp contrast to its traditional reputation of being a cowardly scavenger. In similar fashion recent work by Dr Gus Mills and his colleagues has also revealed a rather different view on the lion's eating habits.

As part of a project aimed at determining the impact of lions on their main prey species, Mills and other researchers followed lions around continuously for several two-week stretches to observe what they killed in precise detail. The generally accepted view, based on data gathered mainly from observations made during daylight hours by park scientists, rangers and visitors, was that lions predominantly killed large prey species like buffalo, zebra and wildebeest, and that a mere 13 per cent of their total prey was impala-sized or smaller. Mills's work disproved this because his findings were that more than 60 per cent of the kills of Kruger lions were of impala or lesser game. The lions made such short work of these kills that often no evidence was left behind to be detected by daytime observers.

Further research in one particular part of the Kruger Park between Lower Sabie and Crocodile Bridge revealed very different patterns of predation by lions on their two favourite large prey species, zebra and blue wildebeest. In the case of wildebeest, bulls, cows and calves were preyed upon in proportion to their ratio of occurrence in the general wildebeest population. This pattern was not followed for zebra, however, where foals made up the majority of the kills. Adult Zebra, it would appear, are considerably more difficult to catch than adult wildebeest. In addition wildebeest in the area are sedentary whereas the zebra are migratory. Wildebeest are available throughout the year, zebra mainly in summer only. For these reasons lions are a major cause of mortality for wildebeest and therefore play an important role in controlling their population numbers, while they have far less impact on zebra populations.

The main research centre for the National Parks Board has traditionally been situated at Skukuza in the Kruger National Park. There are, however, two regional centres in Kimberley and at Rondevlei in the Wilderness National Park.

As a result of the improved focus through the streamlining and rationalization of the Parks Board's structure carried out by Dr Robinson after he assumed the post of chief executive director in 1992 there are now four executive directors responsible for the day-to-day management of the country's national parks. These are:

❑ Dr Salomon Joubert, who is the executive director in charge of the Kruger National Park. This park is a separate management entity because of its large size.
❑ Dr Anthony Hall-Martin, who is in charge of all other parks (Southern Parks).
❑ Ben Mokoatle, the executive director in charge of human resources.
❑ Klasie Havenga, the executive director responsible for the Board's finances.

The Kruger National Park is by far the biggest unit in the system. Not only does it have the largest staff complement but it provides two-thirds of the hut accommodation and half of the camp accommodation available each night throughout South Africa's national park network.

Vaalbos National Park warden Craig Bancroft with white rhinoceros translocated to the park.

This network plays an important role in the country's tourist economy. In the year ending March 1992, the number of visitors totalled 1,09 million. Of these, nearly 700 000 went to the Kruger National Park. The administrative tasks associated with this huge influx of people are enormously complex: in 1992 the Board's central booking offices handled a total of 167 000 applications for accommodation, 114 000 alterations to bookings, 90 000 deposits and 1,11 million enquiries about accommodation. These statistics underline the importance of the network of the national parks to South Africa's tourist industry, an importance that will grow further as the Parks Board implements its expansion plans in response to the need to protect unique ecosystems in the international and national interest and growing eco-tourism business.

THE PARKS IN BRIEF

Reservations for all accommodation and trails can be made at National Parks Board offices in Pretoria and Cape Town.

PRETORIA
P.O. Box 787, 0001 Pretoria
Telephone (012) 343-1991
Fax (012) 343-9770

CAPE TOWN
P.O. Box 7400, 8012 Roggebaai
Telephone (021) 22-2810
Fax (021) 24-6211

KRUGER NATIONAL PARK

GENERAL INFORMATION AND ACCESS

This 1,9 million-ha park is 350 km long, 60 km wide on average and is located along the eastern Transvaal border with Mozambique. The park's headquarters, Skukuza, is situated about 500 km from Johannesburg and can be reached in about six hours' driving. Towns in close proximity to the park are Nelspruit in the south and Phalaborwa in the middle.

Gate times vary throughout the year as follows:

	ENTRANCE GATES OPEN	CAMPS OPEN	ENTRANCE GATES AND CAMPS CLOSE
January	05h30	05h00	18h30
February	05h30	05h30	18h30
March	05h30	05h30	18h00
April	06h00	06h00	17h30
May 1 to Aug 31	06h30	06h30	17h30
September	06h00	06h00	18h00
October	05h30	05h30	18h00
Nov 1 to Dec 31	05h30	04h30	18h30

Comair (telephone (011) 973-2911) offers daily flights from Jan Smuts Airport, Johannesburg, and also operates a number of fly-in safaris. Avis (reservations toll-free telephone 08000 34444) rents cars from Skukuza and meets all Comair flights. National Parks Board offices at Skukuza telephone (01311) 65611.

HABITAT AND CLIMATE

The Kruger Park lies in the Transvaal Lowveld and the climate is subtropical with summer rains (October-March). Annual average rainfall varies from 700 mm in the south of the park to 400 mm in the north. Summers are humid with the daily temperatures usually in the mid-30s Celsius and rising to above 40 °C on occasion. The winter months are pleasantly warm and dry with day temperatures in the mid-20s Celsius. Nights are cool to cold although frost does not usually occur.

The park's habitat is made up of the combination of trees, shrubs and grasses known as bushveld, which can be broken down into four main subdivisions and 35 specific vegetation types or landscapes in total. The most visible change takes place north of the Olifants River where the climate is drier; mopane savanna becomes dominant and baobab trees start to appear.

CAMPS AND FACILITIES

The Kruger Park offers a wide range of hutted accommodation, tents and camping-sites in six bushveld camps, five private camps and 14 rest-camps scattered throughout the park. A number of the camps have 'sponsored accommodation' – various sizes of lodges paid for by private individuals in return for the right to stay there free a certain number of days each year. For the rest of the time the accommodation can be let to the public. A number of the larger camps show films on conservation and wildlife at night.

Bushveld camps are small camps designed to be more rustic and simple than the main rest-camps. They have neither shops, restaurants or petrol stations although they are all near rest-camps which do have such facilities. Electricity for lighting comes from solar panels and stoves, and refrigerators are gas-powered. The various huts in the camps can be hired individually and no day-visitors are allowed. The bushveld camps from south to north are Mbyamiti, Jakkalsbessie, Talamati, Shimuwini, Bateleur and Sirheni.
Private camps are small camps designed to provide total privacy for the group occupying them, and are hired in their entirety. Facilities are similar to the bushveld camps. The private camps from south to north are Malelane, Jock of the Bushveld, Nwanetsi, Roodewal and Boulders.

A modern-style chalet at the Berg-en-dal Rest-camp in the Kruger National Park.

REST-CAMPS

Balule is a small camp 11 km from Olifants Camp providing basic facilities – there is a communal gas fridge but no electricity, shop or restaurant. Light is supplied by paraffin lanterns. No day-visitors allowed. Accommodation consists of six 3-bed huts with public ablution facilities and a communal kitchen with coal stoves. Camping-sites are also provided.

Berg-en-dal offers full conference facilities for a maximum of 200 people and has a restaurant, cafeteria and swimming-pool for residents only. There is a shop, information centre, public telephone and laundry with irons. Petrol and diesel are sold.

❏ Sponsored accommodation – J. le Roux House (6 people) and Rhino House (8 people).
❏ 6-bed cottages with air-conditioning, two bedrooms, bathroom, toilet, lounge/dining-room and fully equipped kitchen. Some cottages are equipped for the handicapped.
❏ 3-bed huts with air-conditioning, shower, toilet and fully equipped kitchen.
❏ Camping-sites with communal kitchens and ablution facilities.

Crocodile Bridge is a small camp on the park's southern boundary with shop, laundry with irons, and a public telephone. Petrol is sold but not diesel. There is no restaurant.

❏ 3-bed huts with veranda, air-conditioning, fully equipped kitchen, shower and toilet. Some are equipped for the handicapped.
❏ Camping-sites with communal kitchens and ablution blocks.

Letaba is situated on the southern banks of the Letaba River. It has a restaurant with a view over the river and an information centre concentrating on elephant. The camp also has Auto-

mobile Association (AA) emergency service and workshops, a self-service cafeteria, shop and laundry with irons. Petrol and diesel are sold.

❏ Sponsored accommodation – Melville House (9 people), Fish Eagle House (8 people).
❏ 6-bed cottages with air-conditioning, two bedrooms and bathroom, toilets and fully equipped kitchen.
❏ 3-bed huts with open verandas, air-conditioning, shower, toilet, refrigerator, two-plate stove and sink. There are no cooking or eating utensils.
❏ 2-bed huts with air-conditioning, bath, toilet and kitchen but no cooking or eating utensils. Some of these huts have facilities for the handicapped.
❏ 4-bed hut with air-conditioning, communal ablution facilities but no cooking facilities.
❏ 4-bed furnished tents with no cooking or eating utensils provided.
❏ Camping- and caravan-sites with communal kitchens and ablution facilities.

Lower Sabie has a restaurant, self-service cafeteria, shop, and laundry with irons. Petrol and diesel are sold.

❏ Sponsored accommodation – Keartland House (7 people), Steenbok Cottage (4 people), Moffet Cottage (4 people).
❏ 5-bed cottages with air-conditioning, two bedrooms, two bathrooms and toilets and fully equipped kitchen with refrigerator.
❏ 2-bed huts with air-conditioning, shower, toilet, refrigerator, and open verandas. Cooking and eating utensils not provided.
❏ 1-, 2-, 3- and 5-bed huts with communal ablution facilities, refrigerator and air-conditioning. Cooking and eating utensils not provided.
❏ Camping-sites are available with communal kitchens and ablution blocks.

Maroela is a small camping area with communal kitchen and ablution facilities but no electricity and no hutted accommodation.

Mopani is 45 km north of Letaba and has a restaurant and ladies' bar, self-service cafeteria, laundry and irons, a swimming-pool for residents only, a shop and an information centre. Petrol and diesel are sold.

❏ Sponsored accommodation – Xanatseni House (8 people).
❏ 6-bed cottages with three bedrooms, bathroom, air conditioning and fully equipped kitchen.
❏ 4-bed luxury huts with fully equipped kitchen, shower and toilet.

An aerial view of Mopani Rest-camp, Kruger National Park.

❏ 4-bed huts with fully equipped kitchen, shower and toilet.
❏ 4-bed huts for handicapped people with fully equipped kitchen, shower and toilet.
❏ There are no camping-sites.

Olifants overlooks the Olifants River and has a restaurant, self-service cafeteria, shop and laundry with irons. Petrol and diesel are sold.
❏ Sponsored accommodation – C.D. Ellis House (8 people) and Cruse House (8 people).
❏ 6-bed cottage with closed veranda, air-conditioning, two bedrooms, bathroom and toilet, and fully equipped kitchen.
❏ 2-bed huts with air-conditioning, toilet and bath, fully equipped kitchen.
❏ 2- and 3-bed huts with open verandas, air-conditioning, shower, toilet, and refrigerator. Cooking and eating utensils not provided. One of the 2-bed huts is equipped for the handicapped.
❏ There are no camping-sites.

Orpen is a small camp situated at the Orpen Gate. It has a shop, and petrol and diesel are sold. Electricity is provided only in the cottages and two of the 2-bed huts. Lanterns are provided in the other huts.
❏ Sponsored accommodation – various 6-bed cottages.
❏ 2- and 3-bed huts with communal ablution facilities, no air-conditioning, fan or refrigerator. Cooking and eating utensils not provided.
❏ 2-bed huts with communal ablution facilities, refrigerator and fan. Cooking and eating utensils not provided. There is no air-conditioning.
❏ There are no camping-sites.

Pretoriuskop is the oldest camp in the Kruger Park with a milder climate than the rest of the park because it is situated on higher ground. It has a restaurant, self-service cafeteria, swimming-pool for residents, shop, and laundry with irons. Petrol and diesel are sold.
❏ Sponsored accommodation – Pierre Joubert House (8 people) and Doherty Bryant Boma (9 people).
❏ 6-bed cottage with air-conditioning, 3 bedrooms, two bathrooms, toilet and fully equipped kitchen.
❏ 6-bed cottages with closed veranda, air-conditioning, 2 bedrooms, one bathroom, toilet, and fully equipped kitchen.
❏ 2-, 3- and 4-bed huts with air-conditioning, shower, toilet and refrigerator. Some 2-bed huts with two-plate stove. Cooking and eating utensils not provided.
❏ 2-, 3-, 5-, and 6-bed huts with refrigerator, air-conditioning and communal ablution facilities. No verandas. Cooking and eating utensils not provided.
❏ 2-bed huts with communal ablution facilities, some with air-conditioning, all without veranda or refrigerator. Cooking and eating utensils not provided.

Olifants Rest-camp commands a superb view of the Olifants River where elephant are frequently seen drinking.

Impala stroll through Pretoriuskop Rest-camp in the Kruger National Park.

❑ Camping-sites are available with communal kitchen and ablution facilities.

Punda Maria is a small camp and the most northerly in the park. It has a restaurant and shop. Petrol and diesel are sold.
❑ 4-bed cottages with air-conditioning, one bedroom, bathroom, toilet, living-room and fully equipped kitchen.
❑ 3-bed huts with air-conditioning, shower, toilet and fully equipped kitchen.
❑ 2-bed huts with air-conditioning, shower, toilet and refrigerator. Cooking and eating utensils not provided.
❑ Camping-sites are available with communal kitchen and ablution facilities.

Satara is situated centrally in the park in an area known for the largest game concentrations. Facilities include AA emergency service for vehicles, restaurant, cafeteria, shop, and laundry with irons. Petrol and diesel are sold.
❑ Sponsored accommodation – Rudy Frankel House (8 people), Stanley House (9 people), Wells House (6 people), nine other 6-bed cottages and one 5-bed cottage.
❑ 2- and 3-bed huts with verandas, air-conditioning, shower, toilet, and refrigerator. Cooking and eating utensils not provided.
❑ 2-bed huts with air-conditioning, facilities for the handicapped, bath, toilet, and refrigerator. Cooking and eating utensils not provided.
❑ Camping-sites are available with communal kitchen and ablution facilities.

Shingwedzi is the second most northerly camp and is situated in prime buffalo and elephant country. It has a restaurant and self-service cafeteria, and there is a swimming-pool for residents, shop, and laundry with irons. Petrol and diesel are sold.
❑ Sponsored accommodation – Rentmeester House (7 people).
❑ One 4-bed cottage with two bedrooms and two bathrooms, toilets, fully equipped kitchen, but no air-conditioning.
❑ Two 2-bed huts with shower, toilet, and fully equipped kitchen, but no air-conditioning.
❑ 5-bed huts with air-conditioning, shower, toilet and refrigerator. Cooking and eating utensils not provided.
❑ 3-bed huts with shower, toilet and refrigerator. No air-conditioning. Cooking and eating utensils not provided.
❑ 3-bed huts with refrigerator but no air-conditioning. Cooking and eating utensils not provided. Communal ablution facilities.
❑ Camping-sites are available with communal kitchen and ablution facilities.

Skukuza is the park's headquarters and has a number of historical sites as well as the Stevenson-Hamilton Memorial Library, an environmental education centre, AA emergency service for vehicles, resident doctor, bank, post office and public telephones, police station, two restaurants, self-service cafeteria, shop, laundry and irons. Petrol and diesel are sold.
❑ Sponsored accommodation – Volkskas House (8 people), Moni House (9 people), Struben Cottage (6 people), Lion Hut (2 people), Nyathi House (8 people), Waterkant House (8 people), Waterkant Cottage (4 people).
❑ 6-bed cottages with air-conditioning, 2 bedrooms, 2 bathrooms and toilets, fully equipped kitchen, lounge/dining room and open verandas.
❑ 6-bed cottages with air-conditioning, 2 bedrooms, one bathroom, toilet, fully equipped kitchen and enclosed verandas as well as one 4-bed cottage with identical specifications.
❑ 2- and 3-bed huts with air-conditioning, shower, toilet, refrigerator and 2-plate stove. Cooking and eating utensils not provided.
❑ 2- and 3-bed huts with air-conditioning, shower, toilet and refrigerator. Cooking and eating utensils not provided.
❑ 2-bed huts with air-conditioning, facilities for the handicapped, bath or shower, toilet and refrigerator. Some have a 2-plate stove. Cooking and eating utensils not provided.
❑ Camping-sites with communal kitchens and ablution facilities.

OTHER ATTRACTIONS AND ACTIVITIES

The park also has 13 picnic-sites in various attractive situations. The road network is excellent and consists of tarred roads on all routes to and between the main camps and good gravel surfaces elsewhere.

Three-night-long wilderness trails are offered in seven parts of the Kruger Park which are closed to the rest of the public.

WARNINGS

Malaria is rife throughout the Kruger and all visitors are advised to begin a course of anti-malaria tablets before they visit the park, no matter what time of year it is.

The bilharzia parasite is present in all streams and dams in the park and visitors should keep out of them.

ADDO ELEPHANT NATIONAL PARK

GENERAL INFORMATION AND ACCESS

Situated some 70 km by road from Port Elizabeth in the eastern Cape, this park covers an area of 12 126 ha and has one rest-camp. The entrance gate is open from 07h00 to 19h00 daily.

Accommodation at Addo Elephant National Park has been upgraded with the completion of these new Cape Dutch style chalets and cottages.

There are daily flights to Port Elizabeth from all the main centres in South Africa and a number of car-hire agencies operate from the airport. The main attraction of the park is the elephant population it protects but buffalo and black rhinoceros are also present.

Addo Elephant National Park office telephone (0426) 400556/7.

HABITAT AND CLIMATE

The park conserves a vegetation type known as Addo Bush, a sub-type of Valley Bushveld. It is a thick and impenetrable low scrub. Because of the Addo's proximity to the coast the climate is warm to hot temperate with an average rainfall of around 450 mm spread throughout the year. Winters are pleasant although rain can occur at any time while, during summer, heat waves can bring the temperature up to 40 °C with high humidity levels.

CAMPS AND FACILITIES

Addo has one rest-camp with restaurant, public telephone and shop. It sells petrol but not diesel. The nearest garage is 15 km away.

❏ 6-bed cottages with 2 air-conditioned bedrooms with bath or shower and toilet, living-room and fully equipped kitchen.

❏ Chalets with kitchen, bathroom, 2 single beds, one double-sleeper and air-conditioning. Fully equipped kitchen. Two of the huts are adapted to accommodate the handicapped.

❏ 2-bed huts with shower, toilet and refrigerator but no air-conditioning. There is a communal kitchen with cooking utensils, crockery and cutlery.

❏ Caravan- and camping-sites are provided with communal ablution, kitchen and braai facilities.

OTHER ATTRACTIONS AND ACTIVITIES

There is a bird-watching hide in the camp and the water-hole in front of the restaurant is illuminated at night. The park also offers nature trails and night game drives. It has a swimming-pool.

AUGRABIES FALLS NATIONAL PARK

GENERAL INFORMATION AND ACCESS

This 82 415-ha park is situated in the northern Cape 120 km from Upington and 840 km from Cape Town. SAA operates daily flights into Upington where several car-rental agencies have branches. The park's entrance gate is open from 07h00 to 19h00 daily. The main attraction of the park is the Augrabies Falls on the Orange River.

Augrabies Falls National Park office telephone 054472 and ask for Augrabies Falls 4.

HABITAT AND CLIMATE

The park lies in an arid, semi-desert region where the average annual rainfall is just over 100 mm. It falls mostly between January and April. Summers can be extremely hot with the temperature frequently rising above 40 ℃. Winter days are pleasant and warm but the nights can be very cold.

CAMPS AND FACILITIES

There is one rest-camp with a shop, restaurant and snack bar. There is also a laundry and ironing room with electric plug-points. Irons are supplied. Petrol is sold but not diesel. The nearest garage is in Kakamas, 40 km distant.

❏ 4-bed cottages with air-conditioning, 2 bedrooms, bathroom, toilet, living-room and open-plan kitchen. Cooking and eating utensils and refrigerator are provided.

Air-conditioned chalets at the Augrabies Falls National Park provide welcome relief from the high summer temperatures.

❑ Chalets with air-conditioning, open-plan kitchen, shower and toilet, two single beds and a double-sleeper.

❑ 2-bed chalets with air-conditioning, kitchen, shower and toilet. Two units are adapted for the handicapped. Cooking and eating utensils and a refrigerator are provided.

❑ Caravan- and camping-sites with communal kitchen and ablution facilities. There is a laundry with irons supplied.

OTHER ATTRACTIONS AND ACTIVITIES

There are a number of viewpoints over the waterfall close to the restaurant, while visitors can drive and view game in the section of the park south of the Orange River. The camp has three swimming-pools for the use of residents only.

The Klipspringer Hiking Trail offers three days of hiking with two overnight huts. Hiking groups are limited to a maximum of 12 persons. The trail is closed from the middle of October to the end of March.

There are also three nature trails at different locations in the park adjacent to the game-drive roads. Each trail takes about an hour to complete.

WARNINGS

A course of anti-malaria pills should be taken before entering the park. Be cautious near the lip of the waterfall and the gorges. There is a protective barrier fence at the waterfall but a number of people over the years have ignored this, ventured too close and fallen to their deaths.

BONTEBOK NATIONAL PARK

GENERAL INFORMATION AND ACCESS

This 2 786-ha park lies in the coastal fynbos area of the southern Cape some 6 km from the town of Swellendam and 238 km from Cape Town. The park was originally established to protect the endangered bontebok.

The entrance gates are open as follows: October 1 to April 30 from 08h00 to 19h00 and May 1 to September 30 from 08h00 to 18h00.

Bontebok National Park office telephone (0291) 42735.

HABITAT AND CLIMATE

The vegetation of the Bontebok Park is known as coastal renos-terveld. This veld-type is one of the major subdivisions of the Cape Floral Kingdom and is renowned for its wealth of plants and flowers, most of which bloom in late winter and spring following the winter rains. The average annual rainfall at the park is just over 500 mm with most of it falling from April to September. The climate is temperate throughout the year with summer maximum temperatures usually in the upper 20s Celsius and sometimes rising over 40 degrees. While the winters are not as bitterly cold as in the interior of the country, a steady drizzle can set in for days at a stretch.

CAMPS AND FACILITIES

The park does not have a hutted rest-camp. Accommodation presently available consists of caravan- and camping-sites using communal ablution blocks which have generated electricity after hours.

The park does offer a limited number of 6-berth caravans equipped with gas stove and refrigerator. Cooking and eating utensils are provided. Each caravan has a side tent.

There is a picnic-site for day-visitors with ablution and braai facilities. Other facilities include a shop and information centre and petrol is available. The nearest garage is 6 km away in Swellendam.

OTHER ATTRACTIONS AND ACTIVITIES

In addition to game-viewing from vehicles there are two short nature trails. Visitors may fish and bathe in the Breede River at the camping-ground.

GOLDEN GATE HIGHLANDS NATIONAL PARK

GENERAL INFORMATION AND ACCESS

This 11 630-ha park is situated in the eastern Orange Free State 338 km from Johannesburg and 54 km from the nearest town, Bethlehem. It conserves scenic, mountainous country on the Lesotho border. There are no entrance gates and reception hours at the Brandwag Rest-camp are from 07h00 to 20h00 daily and at Glen Reenen Rest-camp from 07h30 to 17h00 (Mondays to Saturdays) and on Sundays from 08h00 to 15h00 (summer) and 07h00 to 15h00 (winter). Visitors arriving after hours will find a note on the office door giving their chalet or room number.

Golden Gate Highlands National Park office telephone (0143) 2561471.

HABITAT AND CLIMATE

The area is mountainous with a high rainfall – about 800 mm annually – which falls during the summer months when temperatures are moderate, rarely reaching 30 °C because of the altitude. Winters are very cold with occasional snow, and night temperatures frequently drop below zero.

CAMPS AND FACILITIES

Brandwag has full conference facilities for large (up to 180) and small (maximum 30) groups. Accommodation is in the main building, separate chalets or a youth hostel. The main building has a restaurant, coffee shop, curio shop and ladies' bar.

Glen Reenen Rest-camp at Golden Gate offers traditional hutted accommodation in contrast to the Brandwag hotel complex.

❏ In the main building all rooms are equipped with telephone, radio and TV. Single rooms with a double bed, shower and toilet; double rooms with two single beds, or two double beds, and bathroom and toilet. One suite is available.
❏ 4-bed chalets consisting of living-room with 2 single beds and balcony room with double bed, bathroom, toilet and kitchenette with refrigerator. Cooking and eating utensils are provided. Each chalet has a carport, telephone, radio and TV.
❏ Wilgenhof Youth Hostel – for youth and other groups. Maximum of 86 people.

Glen Reenen has a shop and a picnic-site for day-visitors. Petrol and diesel are sold.
❏ Huts with loft, kitchen, refrigerator, shower and toilet. There are two single beds below and a double bed in the loft. Cooking and eating utensils are provided.
❏ Huts with kitchen nook, shower and toilet, but basically consisting of one room with three single beds. A refrigerator and cooking and eating utensils are provided.
❏ Caravan- and camping-sites with ablution blocks, scullery and braai facilities.

OTHER ACTIVITIES AND ATTRACTIONS

Brandwag offers bowls, tennis, table-tennis, and snooker as well as horse-riding at the Gladstone Stables while ponies can be hired during holidays at Glen Reenen. Glen Reenen also has a natural rock pool in which visitors can swim.

There is a two-day hiking trail – the Rhebok Hiking Trail – with one overnight hut that can take a maximum of 18 people. There are also a number of shorter nature trails lasting from one to five hours.

Guided excursions are offered from Brandwag during holidays while environmental education courses lasting from one to five days can be arranged through the Pretoria booking office.

Film shows and lectures are given every night during school holidays at 20h00 at Brandwag.

WARNINGS

Visitors, and particularly hikers, should remember that mountain weather can change suddenly and it can get very cold at any time of year.

KALAHARI GEMSBOK NATIONAL PARK

GENERAL INFORMATION AND ACCESS

This huge, 960 000-ha park is situated in the northern Cape bordering Namibia and Botswana and offers the visitor the opportunity to discover the extraordinary riches of the semi-desert ecosystem. Entrance to the park is only possible at the main camp, Twee Rivieren. There are two routes in, from Hotazel 340 km to the east and Upington 358 km to the south. Both are gravel roads and, depending on the state of the surface, the journey should take about four hours.

Upington is the nearest major town, as Hotazel is little more than a mining village. There are daily flights by SAA into Upington and vehicles may be hired from a number of car-rental agencies. Visitors to the Nossob and Mata Mata rest-camps have to leave Twee Rivieren by 12 noon to get there by nightfall or they will not be allowed to proceed.

Gate times are as follows:

January-February	06h00 – 19h30
March	06h30 – 19h00
April	07h00 – 18h30
May	07h00 – 18h00
June-July	07h30 – 18h00
August	07h00 – 18h30
September	06h30 – 18h30
October	06h00 – 19h00
November-December	05h30 – 19h30

Light aircraft can land at Twee Rivieren if arrangements are made in advance while cars can be hired from Avis in Twee Rivieren provided also that the booking is made in advance. The nearest garage is at Askham, about 70 km away, but any serious mechanical problem will probably require help from Upington.

Kalahari Gemsbok National Park office telephone 0020 and ask for Gemsbok Park 901.

HABITAT AND CLIMATE

The park is set in red Kalahari sand-dunes with the main roads between the camps running in the (usually) dry river-beds of the Nossob and Auob rivers. Vegetation is sparse and large trees are found only along the river-beds. The climate is dry and harsh with an annual average rainfall of 200 mm which falls mainly between January and April. Summer day temperatures can soar above 40 °C although the nights are usually reasonably cool. In winter the days are usually sunny and warm but the nights are very cold with temperatures dropping well below zero.

CAMPS AND FACILITIES

There are three rest-camps: Twee Rivieren, Nossob and Mata Mata. All three have shops (but only Twee Rivieren stocks fresh meat, eggs and bread) and sell petrol and diesel. Only Twee Rivieren has a restaurant, public telephone and swimming-pool. All accommodation is supplied with cooking and eating utensils but, while all accommodation in Twee Rivieren is air-conditioned, that in the other two camps is not. There are camping-sites at each camp.

Twee Rivieren
- ❑ 3- or 4-bed chalets with kitchen, refrigerator, air-conditioning, shower and toilet.
- ❑ Bedsitter chalets with two single beds and a double-sleeper, kitchen, refrigerator, air-conditioning, shower and toilet.
- ❑ Camping-sites with ablution and kitchen facilities.

Nossob
- ❑ 6-bed cottages with kitchen, refrigerator, two bedrooms, living-room, shower and toilet. No air-conditioning.
- ❑ 3-bed huts with one bedroom, kitchen, refrigerator, shower and toilet. No air-conditioning.
- ❑ 3-bed huts with communal kitchen and ablution facilities. No air-conditioning.
- ❑ Camping-sites with ablution and kitchen facilities.

Twee Rivieren Rest-camp derives its name from its close proximity to the Auob and Nossob rivers.

Mata Mata
- ❑ 6-bed cottages with kitchen, refrigerator, two bedrooms, living-room, shower and toilet. No air-conditioning.
- ❑ 3-bed huts with communal kitchen and ablution facilities. No air-conditioning.
- ❑ Camping-sites with ablution and kitchen facilities.

OTHER ATTRACTIONS AND ACTIVITIES

At present activities in this park are confined to game-viewing from vehicles but four-wheel-drive trails and camel safaris are planned, while game-viewing hides are to be constructed in the near future.

WARNINGS

It is advisable to take a course of anti-malaria tablets in advance of the visit. Scorpions are common and visitors are recommended to wear shoes on summer evenings. In the event of a breakdown the golden rule is to stay with the car. Camp staff will come to look for missing visitors when they fail to report in at their destination by gate-closing time.

KAROO NATIONAL PARK

GENERAL INFORMATION AND ACCESS

This 32 792-ha park is situated in the Great Karoo 12 km from the town of Beaufort West and 471 km from Cape Town. The conserved area consists of typical flat Karoo plains rising into the Nuweveld Mountains which soar about 1 000 m above the rest of the park.

The entrance gates are open from 05h00 to 22h00 daily throughout the year.

Karoo National Park office telephone (0201) 52828/9.

HABITAT AND CLIMATE

This is an arid region with an average annual rainfall of 260 mm falling mainly in the summer months. The habitat consists of Karoo scrub which has no large trees. The climate is harsh with temperatures in the low-lying areas reaching 40 °C in summer although the higher-altitude areas of the park are markedly cooler. Very strong winds may blow at any time of year but particularly in August and September. Winters can be cold with chilly days followed by sub-zero temperatures during the night. Snow sometimes falls on the mountains.

CAMPS AND FACILITIES

There are two rest-camps: the main Karoo Rest-camp and the rustic Mountain View Rest-camp in the mountains.

The Karoo National Park's Cape Dutch style chalets were inspired by the historic architecture of this part of the Cape Province.

Karoo has full conference facilities for a maximum of 70 people, a shop, restaurant, information centre, and a public telephone. There is also a swimming-pool. Petrol and diesel are not available.

❏ 3-bed chalets with air-conditioning, bath, shower, toilet, and kitchenette with refrigerator. Two chalets are adapted for the handicapped. Cooking and eating utensils are provided.

❏ 6-bed family cottages with air-conditioning, two bedrooms, two bathrooms, living-room and kitchenette with refrigerator. One cottage is adapted for the handicapped. Cooking and eating utensils are provided.

❏ Caravan- and camping-sites are available with a scullery, and cooking and ablution facilities.

Mountain View is a somewhat primitive rest-camp in the mountains and can accommodate only 25 people. Beds and mattresses only are provided in the huts. Everything else must be brought in, including firewood; there is a communal ablution block. Book through the Karoo National Park office.

OTHER ATTRACTIONS AND ACTIVITIES

There are two game-viewing roads in the park. The Springbok Hiking Trail is a three-day trail with two overnight huts. A maximum of 12 people per day are permitted to use the trail. It is closed from November to February. There are three nature trails, one concentrating on fossils and another on vegetation. A four-wheel-drive trail was started in 1992.

KNYSNA NATIONAL LAKE AREA

Situated on the coast just west of Plettenberg Bay and some 500 km from Cape Town this park covers the Knysna lagoon only. It takes in no land areas and the Parks Board offers no accommodation here. A wide range of camping and holiday accommodation is available in the town of Knysna and its immediate surroundings. The climate is temperate and cool with rainfall occurring throughout the year and overcast conditions can be expected 50-60 per cent of the time.

There is no control over access to the area below the high-water mark and the lagoon is zoned for activities such as boating, fishing and water-skiing, and for wilderness areas. Knysna National Lake Area office telephone (0445) 22095/22159.

KRANSBERG NATIONAL PARK

Situated some 20 km from Thabazimbi and about 250 km from Johannesburg this park had yet to be proclaimed as this book went to press and was still closed to the public. It is set in the Waterberg range and should be opened to limited numbers of people in the future with the commissioning of picnic sites, caravan camps and four-wheel-drive trails.

Kransberg National Park office telephone (01537) 22229.

MOUNTAIN ZEBRA NATIONAL PARK

GENERAL INFORMATION AND ACCESS

This mountainous 6 536-ha park is set at the eastern border of the Great Karoo 25 km from the town of Cradock and 230 km north of Port Elizabeth. Entrance-gate hours are: October 1 to April 30 from 07h00 to 19h00 and May 1 to September 30 from 07h00 to 18h00. There are daily flights into Port Elizabeth from other South African centres. A number of car-rental agencies operate from the airport.

Mountain Zebra National Park office telephone (0481) 2427/2486.

HABITAT AND CLIMATE

The habitat is Karoo scrub with riverine scrub along a number of the streams draining the mountains. Average rainfall is around 390 mm annually with most falling in summer. The summers are warm with temperatures in the mid- to upper 30s Celsius while the winters are sunny with very cold nights. Snow sometimes falls on the mountains.

CAMPS AND FACILITIES

Mountain Zebra National Park has one rest-camp which has a shop where groceries may be bought, and a restaurant. There are washing and ironing facilities and both petrol and diesel are sold. The camp has full conference facilities for a maximum of 60 people. There are several picnic areas and a swimming-pool for residents.

Doornhoek Guest Cottage in the Mountain Zebra National Park is a restored national monument.

❏ 4-bed cottages with air-conditioning, two bedrooms, bathroom, toilet, living-room and kitchen with refrigerator. Cooking and eating utensils are provided.

❏ Caravan- and camping-sites are provided with ablution and braai facilities and a scullery.

❏ Doornhoek Guest Cottage is situated about 2 km from the main rest-camp and is a restored Victorian farmhouse which is a proclaimed national monument. It accommodates six people and has three bedrooms, three bathrooms, toilets, fully equipped kitchen and a charming Victorian living-room.

OTHER ATTRACTIONS AND ACTIVITIES

There are several game-viewing drives. The Mountain Zebra Hiking Trail lasts three days and two nights and the maximum group size is 12 people. There are several nature trails of different durations, from one hour to a whole day. Horse-riding can be arranged if booked in advance at the office.

RICHTERSVELD NATIONAL PARK

This park in the remote far north of Namaqualand was not yet officially open to the public as this book went to press. The nearest towns are Alexander Bay about 100 km distant and Springbok about 350 km away. The park is still being developed and a series of trails, camping sites and small, simple rest-camps should be opened up shortly.

Richtersveld National Park office telephone (0256) 506.

TANKWA KAROO NATIONAL PARK

This park is situated in the western Karoo about 70 km south of Calvinia and some 250 km from Cape Town. It is completely closed to the public at the moment and limited facilities will be provided in the near future, giving visitors access to the vastness, tranquility and remoteness of the area.

TSITSIKAMMA NATIONAL PARK

GENERAL INFORMATION AND ACCESS

The park takes in 80 km of coastline east of Plettenberg Bay which is 68 km from the main rest-camp. The nearest major city is Port Elizabeth which is 141 km to the east, while Cape Town is 615 km to the west.

The entrance gates are open from 05h00 to 21h30 throughout the year and, after office hours close at 18h00, the keys of reserved accommodation can be collected at the entrance gate up to 21h30.

Tsitsikamma National Park office telephone (04237) 607/651.

HABITAT AND CLIMATE

The habitat is rocky coastline rising sharply through Afro-montane forests to fynbos on the higher ground. The De Vasselot section of the park takes in several stretches of sandy beach and large areas of forest and fynbos. Many handsome tree species can be seen in the dense woodlands. The climate is cool temperate with an average annual rainfall of 1 200 mm. This falls throughout the year although the wettest months are May and October and the driest months June and July.

CAMPS AND FACILITIES

The park has two rest-camps: Storms River Mouth and De Vasselot.

The chalets at the Storms River Mouth Rest-camp in the Tsitsikamma National Park have all been built as close as is prudently possible to the sea.

Storms River Mouth has conference facilities for a maximum of 40 people, a shop, restaurant, laundry, and public telephone. Petrol and diesel are not available.

❑ Beach cottages each with bathroom, toilet, living-room, and kitchen with refrigerator, stove and cooking and eating utensils. The cottages are available in three types of configuration to take 4, 7 or 8 people.

❑ 'Oceanettes' which are essentially a block of flats with two kinds of accommodation:

two bedrooms each with two beds, bathroom, toilet, living-room and fully-equipped kitchen with refrigerator and cooking and eating utensils provided, or,

one room with three single beds, bathroom and toilet, open plan kitchen with refrigerator, stove and cooking and eating utensils.

❑ Caravan-sites for 70 caravans and camping-sites for 30 tents. Ablution and braai facilities are provided.

De Vasselot is a smaller camp situated at Nature's Valley, 40 km west of Storms River Mouth. It has 45 sites for tents and caravans, with ablution and braai facilities but no electricity. There is a telephone but no shop.

OTHER ATTRACTIONS AND ACTIVITIES

At Storms River Mouth visitors may angle, scuba-dive and snorkel, bathe in the sea or in the swimming-pool. There are demarcated underwater trails which may be undertaken by qualified scuba divers. There are four short nature trails as well as the famous 42-km-long Otter Trail which takes four days to complete. Bookings for the Otter Trail should be made at least a year in advance. Guided outings can be arranged and film shows are given in evenings during the school holidays.

At De Vasselot there is a network of six nature trails, while there is board-sailing, rowing, sailing and canoeing on the river, and bathing in the sea at Nature's Valley.

VAALBOS NATIONAL PARK

This park, situated on the Vaal River about 60 km west of Kimberley, is at present closed to the public. It is expected, however, that a picnic/day visitor's site and a road system will soon be available. Other facilities for this park will include a tree-house overlooking a waterhole where small groups may stay overnight.

Vaalbos National Park office telephone (05352) 9012.

WEST COAST NATIONAL PARK

GENERAL INFORMATION AND ACCESS

The park stretches along the Cape's west coast from Yzerfontein to Langebaan which is 122 km north of Cape Town. Although there is no entrance gate at present, it will be instituted in the near future. Reception hours at Langebaan Lodge are from 07h00 to 21h30 throughout the year. West Coast National Park office telephone (02287) 22144.

HABITAT AND CLIMATE

The park takes in strandveld (a veld-type subdivision of the Fynbos Biome), salt-marshes, the Langebaan Lagoon, four islands and the Atlantic seashore. It lies in the winter-rainfall belt (May to August) when a series of cold fronts from the Atlantic can produce a succession of rainy days. The climate is temperate with cool wet winters and hot dry summers. Strong winds are relatively frequent and can be expected at any time of year.

CAMPS AND FACILITIES

Langebaan Lodge, situated in the village of Langebaan, is a hotel offering a variety of rooms. All have a bathroom, toilet, telephone and television set. The rooms have no cooking facilities but there is a coffee-/tea-maker in each.

The hotel has conference facilities, a restaurant, coffee bar, ladies' bar and curio shop.

OTHER ATTRACTIONS AND ACTIVITIES

Although the area is famous for its spring flowers, its bird-watching is excellent throughout the year, particularly in the summer when the migrant waders from the northern hemisphere are present. The Postberg section of the park is only open to the public during the flower season from August to September.

The Langebaan Lodge and National Parks Board offices are ideally situated on the edge of the lagoon in the West Coast National Park.

Sailing conditions are good and the lagoon is an excellent venue for board-sailing in the zoned areas. Bathing beaches are at Langebaan and Kraalbaai while angling is allowed in the general-usage zone of the lagoon and in Saldanha Bay.

There is an educational centre at the historical Geelbek homestead which has been restored. Groups may stay here for environmental courses and trails, for which bookings are made at Langebaan Lodge.

WARNINGS

During tidal changes strong currents develop on both sides of Schaapen Island which may be dangerous for boats, board-sailors and bathers.

WILDERNESS NATIONAL PARK

GENERAL INFORMATION AND ACCESS

This national park includes a series of lakes and coastal areas along the Indian Ocean and is situated about 15 km east of the town of George and 451 km east of Cape Town. There is no entrance gate and reception office hours are from 08h00 to 13h00 and 14h00 to 17h00 except during the December/January school holidays when the office is open from 07h00 to 20h00 daily. The offices are about 3 km east of the village of Wilderness.

Wilderness National Park office telephone (0441) 91197.

HABITAT AND CLIMATE

The park conserves several lakes, estuaries and large tracts of reed-bed habitat as well as sandy shorelines and forested slopes inland. The climate is cool temperate with rain all year round, although September and October have the highest rainfall with 130 mm a month on average, while February is usually the driest with about 75 mm. Expect overcast weather for 40 to 50 per cent of the year.

CAMPS AND FACILITIES

There are two rest-camps.

Wilderness has a shop and public telephone but no restaurant.
- ❏ 4-bed log cabins with two bedrooms, bathroom, toilet, living-room and fully equipped kitchen. Cooking and eating utensils are provided. There is covered parking for one vehicle.
- ❏ 6-bed cottages with two bedrooms, a living-room, bathroom, toilet, and a fully equipped kitchen with cooking and eating utensils provided. There is covered parking for one vehicle.
- ❏ 6-berth 'chalavan' which consists of a caravan under a roof and with an anteroom with gas stove and refrigerator. Cooking and eating utensils are provided.

The log cabins at Wilderness are raised on stilts to command a view of the surrounding estuary and reedbeds.

- ❏ Caravan- and camping-sites are provided with ablution and braai facilities and a laundry and dryer.

Ebb and Flow is situated adjacent to the Wilderness Rest-camp. It has no shop, restaurant or public telephone.
- ❏ 2-bed huts with shower and toilet, washbasin, refrigerator, and 2-plate stove. Cooking and eating utensils are provided.
- ❏ 2-bed huts with communal ablution facilities, washbasin, refrigerator, and 2-plate stove. Cooking and eating utensils are provided.
- ❏ Caravan- and camping-sites are provided with ablution and braai facilities.

OTHER ATTRACTIONS AND ACTIVITIES

Bathing, rowing and boating on the beach at Wilderness, the Touw River lagoon and the lakes. Water-skiing is only allowed on Island Lake and Swartvlei. Pedal-boats, rowing-boats and canoes can be hired at the rest-camp. There is a network of day nature trails covering the forests, beach and lakes. There are two bird-watching hides at Langvlei and Rondevlei.

ZUURBERG NATIONAL PARK

Situated about 90 km from Port Elizabeth, this national park covers 35 000 ha of mainly mountain fynbos habitat and is still under development with only limited access to the public.

Two-day hiking trails are available, based on the eastern side of the park. They are reached from Addo. The only accommodation available at this stage is a fully equipped 6-bed guest-house at Kaboega on the western side of the park reached from the village of Kirkwood. Bookings can be made through the office at Addo Elephant National Park. The telephone number at Addo is (0426) 400556.

INDEX

PHOTOGRAPHIC CREDITS

All photographs in this book supplied by APBL (Anthony Bannister Photo Library), PO Box 11, Lanseria, 1748 South Africa. Tel: (011) 701-3000 Fax (011) 701-3003

The following photographs are by ABPL photographers other than Anthony Bannister. Picture credits for each page read from top to bottom. Where the tops of two or more pictures are on the same level, credits read from left to right.

1 Clem Haagner
4 Nigel Dennis
8 Clem Haagner
10 Wayne Saunders
11 Nigel Dennis
14 Nigel Dennis
16 Eric Reisinger
17 Peter Pickford
18 Brendan Ryan (African Images)
27 Clem Haagner
28 Brendan Ryan (African Images)
30 Right: Philip Richardson
31 Nigel Dennis, Joan Ryder
39 Right: Clem Haagner

41 Joan Ryder
42 Philip Richardson
43 Clem Haagner
44 Terry Carew
45 Clem Haagner
46 Left: Hein van Hörsten
48 Nigel Dennis
49 Clem Haagner
50 Clem Haagner
51 Clem Haagner
52 Left: Clem Haagner
54 Nigel Dennis
55 Nigel Dennis
57 Roger de la Harpe
60 Clem Haagner, Clem Haagner
62 Right: Nigel Dennis
65 Karl Switak
68 Phillip Richardson
69 Nigel Dennis, Hein von Hörsten
71 Rudi van Aarde
77 Right: Hein von Hörsten
83 Left: Nigel Dennis
84 Malcom Funston
87 Peter Pickford
88 Clem Haagner
89 Clem Haagner
90 Lorna Stanton
91 Peter Pickford
92 Roger de la Harpe
93 Clem Haagner

94 Malcom Funston
95 Clem Haagner
96 Peter Pickford, Richard du Toit
99 Richard du Toit
101 Nigel Dennis
102 Lorna Stanton
103 Robert Nunnington
104 Wayne Saunders
105 Nigel Dennis
106 Roger de la Harpe
107 Nigel Dennis
109 Roger de la Harpe
110 Hein von Hörsten
111 Nigel Dennis
115 Peter Pickford
129 Hein von Hörsten
135 Roger de la Harpe
144 Lorna Stanton
146 Herman Potgieter
150 Nigel Dennis, Nigel Dennis
151 Hein von Hörsten, Peet Joubert
153 Clem Haagner
155 Lorna Stanton
156 Lorna Stanton
157 Daryl Balfour
160 Johan van Jaarsveld
169 Roger de la Harpe
170 Left: below: Lesley Hay, Reg Gosh
172 Below: National Parks Board
179 Lorna Stanton

Front cover: Joan Ryder
Spine: Lorna Stanton
Back cover: top left Paul Funston;
top right, middle left and right, and bottom left Anthony Bannister;
bottom right Joan Ryder